W9-AVM-222

PEOPLE UNLIKE US

PEOPLE UNLIKE US

jeremy j. millett

an imprint of Prometheus Books
59 John Glenn Drive, Amherst, New York 14228-2119

Published 2008 by Humanity Books, an imprint of Prometheus Books

Inquiries should be addressed to
Humanity Books
59 John Glenn Drive
Amherst, New York 14228–2119
VOICE: 716–691–0133, ext. 210
FAX: 716–691–0137

12 11 10 09 08 5 4 3 2 1

Library of Congress Cataloging-in-Publication Data

People unlike us / Jeremy J. Millett.
p. cm.
Includes bibliographical references and index.
ISBN 978–1–59102–637–2 (hardcover)
1. Social evolution. 2. Social history. 3. Social classes—History. 4. Political culture—History. I. Millett, Jeremy J., 1938–

HM626.P423 2008
303.4—dc22

2008007643

Printed in the United States of America on acid-free paper

CONTENTS

PART 2: THE PRESENT

PART 3: THE FUTURE

Note that the divisions are very rough and arbitrary. Thus while the greatest impact of feudal people was in the past, there remain many feudal people today. In the same way, while we have a few relatively autonomous people today, their great impact will be in the future. Evolution is a continual set of experiments, so some of us lag behind while others advance far beyond what operates upon the majority at any given time.

ACKNOWLEDGMENTS

As usual in cases like this, the author admits all the help obtained from a myriad of sources yet admits as well that he is responsible for the finished product, for good or ill, or, as generally happens, for mixedly both.

Much of the inspiration, particularly for the "present" and "future" parts, comes from Rutgers University, where I had the opportunity to teach many different courses and therefore the chance to think in new and different ways. In particular I had the advantage of teaching two seminars developing the substance of the book, working with a number of very able students who were interested and got quite involved. My thanks to Aaron Werschulz and Dara Elaine Trust, whose thoughts improved my range of thinking greatly, and from whom I have borrowed in the latter chapters. Also Jeremy Lackey, for useful research, and Alexsendr Finkelshtyn, Troy Mulet, and Robert

Ford, in whose useful disagreements with many of these ideas here I found a way to check my thinking. Thanks also to Audrey Boyd, who played a key role in setting up the seminars, and to two understanding and helpful department chairs, Roy Licklider and Ed Rhodes.

My wife, DeAnna, has proven the value of genuine domestic partnership by her skills at proofreading and indexing; Paul McCartney of Towson University has read much of the manuscript and offered a plethora of good ideas; and of course the people at Prometheus, from Steven L. Mitchell and Peggy Deemer on, are doing a fine job with every aspect of preparing the book for its readers.

INTRODUCTION

Thoughts on the Past, Present, and Potential Future of Our Species

Now the struggle between the old, barbarous past and the new, possible future involves the whole human race.[1]
Robert LeFevre

When I was thinking about a quote to head this introduction, the one above jumped out at me. On the excessively hot day in July as I was writing, events around us appeared to reflect this statement and make its value more apparent and its concern more urgent. The war of every tribe against every other tribe continued without letup in Iraq; Hizbollah in Lebanon was trying to kill as many innocent citizens in Israel as possible (they managed to fire mortar shells quite as indiscriminately on Arabs living in Israel as on Jews), making one suppose that ordinary citizens, not soldiers, were the targets. The Israelis were shelling in return. In fact, like

Hizbollah, they killed many others than their targets—reports in the paper one day said Canadians on vacation had been killed, and I have before me a report by an American student in Lebanon, who was huddled with others at the American University, hoping to stay alive. She could not leave: the US embassy was closed and the airport close by had been bombed by the Israelis, despite the fact that it was not a Hizbollah hangout. On a nearby front, the Israelis again invaded Gaza, allegedly going after the Hamas terrorists who sent mortar fire their way; the Israelis started by taking out the power plants so no Palestinian could have water (electricity is needed for pumps), light, or transportation. It is obvious that the Israelis killed many of those they found, just as Hamas or Hizbollah had done.[2] There are no sides to take here, no right and no wrong from our viewpoint, no atrocity for the participants too unjust to commit. They do not believe it is an atrocity to torture and kill those who are not of their tribe. Elsewhere, the slaughter of innocents in Darfur continues, young girls are killed by their families for tribal reasons of honor in Turkey, and in altogether too many places around our globe, similar horrors exist. When we go to sleep, it is not in peace but apparently out of exhaustion. The old barbarous past seems to make the headlines every day. Is there a possible future that does not involve this butchery? I believe so.

If the statements above drive you to pessimism, you need to remember also that Charles Taylor, the principal warlord of Liberia, was in prison awaiting trial on charges of crimes against humanity. This in turn should call to mind that international courts are now in operation and rendering decisions for the first time in human history. Rules for less inhumane warfare, like the Geneva Conventions, are and remain our standards, however much we fall short of living up to them, and regardless of a minority dislike of those limits. Vast China, home not so long ago to purges and murders of millions of innocents, is now in a race to produce consumer goods—occasionally in an environmentally sound way—and see to the growth of its people. Russia, now home to great corruption and occasional murder—and having a state that now accepts and abides by the rulings of the European court[3]—just a few years ago was home to the extermination of millions by the secret police and the communist party. Europe has passed by its bloody history of internal fighting and colonizing the world and has become a peaceful continent united under the general auspices of the European Union. In the United States, huge, hateful organizations like the KKK have been replaced by minuscule hateful groups, and a ruling party, which has tried to operate in secret and make war simply on suspicion that someone is thinking naughty thoughts about us, was brought to heel by an electorate that now gives the leader of that party an approval rating of less than one-third of the population. And finally, in Israel, Lebanon, and even in Palestine there are groups and individuals striving for peace and tolerance of others even in the midst of carnage.[4]

One can find this kind of progress reported almost on a day-to-day basis. As examples, on the last day of July 2006, the Congo was reported as holding a real election, the first in almost fifty years. There were not many problems connected with it, and United Nations supervisors (another achievement of our times) were there to watch for any wrongful actions.[5] On another continent, it was reported that black fever may soon be eradicated. A small nongovernmental organization, a private charity based in San Francisco, had uncovered a simple, very inexpensive drug with almost no side effects, gotten it though all the government red tape, and is now supplying it to people in India. The disease is found in other areas in Asia, Africa, and South America, and apparently the drug will be used there as well. Black fever is the second-most deadly parasitic killer on the planet.[6]

Yet the evil we have with us does not seem to be hurrying to leave. What causes it and what will, in time, eliminate it? I hope I can help us find the answers, and this book is my personal testament to my efforts and my odyssey in search of reasonable hopes. Time and the remarkably rapid evolutionary changes that are now going on in our species are my guides here.[7]

Perhaps a bit of my background will help clarify what I am developing here. For years and years as I taught political philosophy, I treated humanity as if we had some overarching human nature, which led nicely and inevitably to the social contract. I was a good, or at least an adequate, Lockean. From time to time, however, I was troubled. Many years before, I had read Albert Jay Nock, the great American libertarian essayist, and I had been deeply moved by and convinced of the importance of many of the things he had to say. As time went by, I began to use the writings of Franz Oppenheimer, the German American sociologist inspired by Nock,[8] to provide a richer framework for what I had to tell my students. After some years, that background moved to the forefront of my thinking, as many of my more recent students will testify.

The time came when I decided to finally confront a key point from Nock, a point he had borrowed from a man named Cram. Nock had improved upon it substantially, but it still troubled me.[9] Why do some people act so differently from others? A good social contractarian should find we have an underlying human nature—it is the basis upon which any idea of a social contract is built. And while I had long since given up the notion that we actually had a social contract in this country—Lysander Spooner[10] had convinced me of that by pointing out how the simple necessities of agreeing to a contract had never been present except for those who actually signed the Constitution—was I ready to throw over the very concept of the social contract, because the idea of a stable, unchanging human nature was wrong? Yes, I was. And as the story of our development unfolds, I may convince you that I was right in doing so.

Anyway, I rejected the idea that all of us, from Attila to Gandhi, from Sophocles to Hitler, have the same nature. I accepted the position which argues that we are indeed quite different, one from other; hence, the title of this book. Why are we different? Because of our *individual, changing, genetic natures*. Then my need to understand human differences met up with newspaper reports[11] about the continuing process of evolution in people's brains, and then I read about how indeed our basic building blocks, our genes, change rapidly, not just by some slow, generation over generation adaptation to environment, but in remarkably speedy fashion, that "random mutation" is in fact "ubiquitous mutation."[12] Thus, I later also discovered that I could deal with what Deirdre McCloskey called the "chief question of the social sciences,"[13] which appears to be the question as to why we differ so greatly from our relatively recent ancestors. So I went back to Nock and said, yes, Nock, you are right, but we can understand human difference not by merely postulating as you did, that most folks are "merely the sub-human raw material out of which the occasional human is produced"[14] by some mysterious process, but we can make the idea less mysterious, and can thus reasonably accept the idea what while we are all human, some people are more developed, in genetic and moral ways, than others. The workings of evolution upon our species seem to make it so. Progress is possible, and if we do not blow ourselves up or totally poison our planet, we will progress. If we cannot *not* become more human, since we are already human and cannot deprive someone of that even if we don't like him or her, we will nevertheless become more humane, because our environment and our survival will require it of us, and our choices, made within the limits and framework of our evolving brains, will do the same. Our whole story as a species is a study of adaptability in an incredibly wide variety of climates, social conditions, and political circumstances. I believe it gives us hope for our future. As biologist Garrett Hardin pointed out long ago, say what you will about evolution up to now, but you had better understand that it is there, and you had "better believe" it will continue to work, if you also want to believe our species has a future.[15]

Let me sketch out the story of those in the past who are people unlike us (although there was an occasional individual who was very much like us), as well as a study of our contemporaries, with a look ahead to the future generations who will also be people unlike us, and I believe, much better adapted to dealing with each other and the universe around them than we are in our transitional period to civilized conduct.

So, what I am doing, and what I believe to be the key part of my work, is developing the hypothesis that indeed the brain is evolving, that this is shown by the changing ways many of us act, and that a good part of our evolution is fitting us more satisfactorily to the requirements of our new, changed existence. Old tribal man with his brutalities was no doubt well suited for our early development, but our circumstances have changed. We need people who are more flexible, both more individualistic and more cooperative at the same time. Of course, we have always had a few of the more developed folk with us, and now, while we can still see tribal peoples around us, we can see many others, the "changelings," as well. Granted, we are not all one way or another, we do not neatly fit into anyone's pigeonholes; but there are still basic types, and the percentages of these are being modified over a relatively short period of time. So, when you look at the table of contents, you can easily understand that the divisions into past, present, and future are very rough. Our range of choices is now also wider, our abilities to make more choices grows, and the choices we in fact make are more likely to be more closely aligned with our developing needs as individuals and a species.

Here is a bit of our story, with all the hopes we have for a better future, with due apologies to all the authors I borrow from, combine, and elaborate upon. Here also you will find my plea to all that we act upon the very best that is within us individually. We can stretch and grow, and we need to do more of it. Our species needs this growth, and we as individuals need it as well.

NOTES

1. Robert LeFevre. *The Nature of Man and His Government* (Caldwell, ID: Caxton Printers, 1959), p. 11.

2. It may not be necessary to show that the Muslim extremist groups hate Israelis and will gladly kill them. But do the Israelis have a similar disregard for others? Israeli prime minister Olmert, in discussing the war, said that while he was "sorry" for the deaths of others, he regarded the "lives, security and well-being of the residents of Sderot" as more important. It is unclear how one balances this: Are the lives of ten Palestinians or Lebanese or Canadians worth more or less than one Israeli? How does the "well-being" of one Israeli compare to the life of a Palestinian? We might even ask if, in the long run, Israeli lives are really protected by the slaughter of innocents. But this is a judgment from the standpoint of city or global people, not tribal ones. See the story from June 23, 2006, and accessed on that date, at http://news.independent.co.uk/world/middleeast/article1095841.ece.

3. Peter Finn, "Europe's Long Legal Tether on Russia," *Washington Post*, October 23, 2006, p. A1.

4. Note Graham Usher, "Revenge Suicide Bombing Fails to Derail Israel's Peace Movement," *Observer*, February 17, 2002. One might wish to check some useful Web sites: http://batshalom.org, http://www.palsolidarity.org or http://zope.gush.shalom.org (accessed July 4, 2007).

5. Jeffrey Gettleman, "Congo Votes in Its First Multiparty Election in 46 Years," *New York Times*, July 31, 2006, p. A7.

6. Stephanie Strom, "A Small Charity Takes the Reins in Fighting a Neglected Disease," *New York Times*, July 1, 2006, p. A1.

7. See the remarkable work by Thomas Landon Thorson, *Biopolitics* (New York: Holt, Rinehart and Winston, 1970). It is mostly an examination of academic political science literature, but when he speaks of his own position, he points out that one cannot understand politics without looking at the "time dimension" (p. 3). And his point that all higher animal development—including us—involves "group living, defense of territory and intragroup dominance" (p. 124) opens the door though which I am going here.

8. Franz Oppenheimer, *The State*, translated by John Gitterman, introduced by C. Hamilton. This edition came first to hand, and its pagination is the same as the original. It is now available in a variety of editions. It can also be found online at http://www.franz-oppenheimer.de, last accessed June 14, 2007. The cautious reader should note that a great deal of the material in my first few chapters is enhanced by Oppenheimer's analysis, added to my almost four decades of teaching and my development of other sources and, in many cases, somewhat different analysis. The entire genetic approach, in particular, is not present in Oppenheimer.

9. Albert Jay Nock, *Memoirs of a Superfluous Man* (New York: Harper & Brothers, 1943). See also his *Our Enemy the State* (Caldwell, ID: Caxton Printers, 1959).

10. Lysander Spooner, *No Treason: The Constitution of No Authority* (Boston: self-published, 1870), reprinted numerous times. It can be found online at http//www.fourmilab.ch/etexts/www/Notreason/Notreason.html (accessed January 17, 200).

11. See, for example, Nicholas Wade, "Brain May Still Be Evolving, Studies Hint," *New York Times*, September 9, 2005, p. A14.

12. Glenn McGee, *Beyond Genetics* (New York: William Morrow, 2003). See the discussion on pp. 52–61.

13. Deirdre N. McCloskey, *The Bourgeois Virtues* (Chicago: University of Chicago Press, 2006), p. 3.

14. Nock, *Memoirs*, p. 137.

15. Garrett Hardin, *Nature and Man's Fate* (New York: Rinehart and Company, 1950), p. viii.

Part 1

THE PAST

CHAPTER 1

TRIBAL PEOPLE

Mankind consisted of
128 people.[1]
William Tenn

In the beginning, men were mere parts of tribes. They were appendages to the body that was the people. Tribes were simply coalitions of males for the purpose of using violence against others in order to survive[2]—women and children were not really part of the tribe. Of course, there were rare exceptions. The Sarmatian women fought on horseback alongside the men, and apparently it was the tribal custom that a woman could not get married until she had killed someone in battle.[3] But the usual custom by far was that a male child might hope to grow up and become a tribal member; female children had no such hope. They would be traded to other tribes and given to one of their males. Or, if their males did not sufficiently defend their tribal prerogatives, the men would be killed, women of child-bearing age taken and raped, every female older than prime bearing age eliminated, and small children, male and female alike, would be killed quickly and easily.

This is the way we passed our early period of evolution. It can still be seen in primitive tribes that have recently been studied, such as the Yanomamo in Venezuela, where they fight bitterly and fiercely with other like tribes.[4] Why? Because if you are male and a successful killer, you will get more women—you take the women of those you have killed and have many more children. If you don't kill, you are killed and you have no offspring. There are no primitive peaceful tribes of hunter-gatherers unless there are no other tribes around. Among other primitives in Latin America, there are also the Waorani, among whom the rate of violent death was calculated at an amazing—to us—60 percent.[5] As careful students of tribes chillingly tell us, "Better fighters tend to have more babies, that's the simple, stupid, selfish logic of sexual selection."[6] This is why there are so many descendants of Attila around and why many males enjoy claiming to be one of those descendants.[7]

Finally the anthropologists point out that a number of animals use their savvy, or intelligence, to achieve their goals. And evolution is clearly the reason why "problem solving and learning (and the variable behavior that these abilities create)" has evolved.[8] It is useful to the problem-solving bird that gets a clam and flies up and drops the clam on a rock to break it open. Humans are far more able to solve problems, so if primitive tribal males needed to kill for the benefit of the tribe, they learned to kill and to enjoy it. If a complete lack of remorse over the death of others was useful to the survivors, that was our evolutionary stance at that point.

Randolph Bourne, writing at the time of the First World War (and in strong opposition to it) finds similarities between the most primitive and violent tribal behavior and the way that states behaved even in the twentieth century. He says, "The state is the organization of the herd to act offensively or defensively against another herd similarly organized."[9] It is certainly true that he is being harsh on human herds, and while the usage is polemically sound, it misses as analysis when he goes on to claim that the difference between these tribes or herds and the modern state is one of degree of sophistication and variety of organization, and not of kind.[10] I am going to argue quite differently, that the difference between tribe and state is much greater than he will admit, although it is true that in time of war even the most modern states are likely to demonstrate much herdlike or tribelike behavior (we call it *patriotism*). Of course, the earliest states that have directly and immediately risen out of tribal conquests differ a good deal less from simple tribes than do more advanced states.

On the other hand, an enthusiastic early lover of the state, Plato, gives a more neutral application to the term when he speaks of a statesman exercising the "art of man-herding,"[11] and he certainly finds no complaint to raise against it, at least not when it is done according to his specifications.

Some of this may simply be playing "the Venereal Game," after the curious and compelling work by James Lipton.[12] The argument there is that a group name should reflect something key about the group. Accepting that, should we use *tribe, herd,* or *gang* for our early selves? Or is the "male exclusionary" band the best?[13] I opt for the first, as the opportunity for growth and change in tribes, and the ability of people to ultimately transform themselves may be reflected in this name for our early organization. *Herd* and *gang* may be overly pessimistic, nonhuman names.

But to the tribe: Many people have studied the peoples in this condition, and while they seem to mostly agree with Franz Oppenheimer that the state is our first real institution evolved from tribal culture and behavior, they talk about it in very different fashion. For example, some authors talk about this kind of society existing at various times, from precapitalist Europe to America Indians, from African tribes to the tribes of China. These writers point out some of the similarities, from being very slow to change, to "dependence on kin organization" and the unity within each tribe.[14] This unity is described by some sociologists as "tradition-directed"; and as such, they say that a tribal person "hardly thinks of himself as an individual."[15] James Bowman suggests that while it is true that tribal people are like that, he ascribes it to their being part of an "honor group"[16] (this may be a wonderful euphemism for rampaging male killers) and that members of such a group above all owe obedience to their leaders, and believe in the "eye for an eye" theory—something Bowman finds people generally must learn on the playground, as part of human nature.[17] Tribespeople find no value in others not of their tribe.[18] This is why, for example, while lying to people outside the tribe is acceptable and indeed encouraged if it appears to benefit the tribe in the slightest, it is not acceptable to lie within your tribe or honor group. One can lie to lesser creatures, but not to fellow humans.

It is surprising how much of the earliest part of our history is known in general outline: where we came from; how in the struggle of tribe against tribe, some tribes were pushed out of their homeland and moved into virgin territory; how some more adventurous tribes left to go exploring, settled in some new, productive, and interesting place and never came back.[19] But since my explorations with you will provide more of a sociological than historical/anthropological approach, we need consider only the tribe itself, and use various examples merely to demonstrate specific points of concern or interest. Generally speaking, by the time we are first interested in our ancestors, they were divided into two kinds of tribes: the hunting tribes and the farming tribes. These tribes were quite rudimentary, at the start being little more than extended families. If they hunted, they had sticks, perhaps with a rock tied to the end with a vine, or primitive spears or swords, also wooden, which were given an edge with a sharp rock. They were nomadic, going hither and yon following the game. They had a leader, perhaps the young male who was the best tracker or best killer of the game that was available. He had no authority beyond his skills to find and/or provide meat. He was always on his mettle, not only ready to be tried by any up-and-coming tribal member who might think he was better, but he was also at the mercy of the climate, like any leader of any pack of mammals. If rain washed out the trail of the game, the tribe went hungry, and if it happened too often, he was killed and replaced because he was "bad luck." If someone bested him in combat (and this tended to happen as he aged), he died. A tribe has little tolerance for a deposed leader. After all, it is the whole tribe that matters; any member is dispensable and if he is not helping the tribe—if he is too old to track the game properly or if he has simply lost his luck or skill at finding animals to eat—then best he be killed promptly. The logic of this is clear. Tribal survival is at stake, and the evolution of humans at that point was totally directed to group survival against other groups and a hostile environment. At most, with more advanced tribal groupings, the tribe would simply leave their ex-leader behind to starve by himself.

Every tribe had its rituals, its habits and customs. Its attitude was essentially that it believed itself, in some way, to be a chosen people, beloved of the tribal deities or, like the quote at the head of the chapter, the only true people.

Agricultural tribes developed much later. They had little more than the hunters—a few sticks to dig holes, a careful knowledge of natural fertilizers, crude huts to live in, a leader who would decide the simple questions as to who farmed where on the communal plot and who led the rituals.

No tribe had the sort of leaders who could be described as politicians. Jared Diamond points out that leadership at this time is not established in any institutional fashion.[20] Thus, leaders of a hunting tribe may have had a slightly better set of skins to wear, leaders of the peasants may have had a little more elaborate headpiece, but they were all dependent on their abilities to bring down deer or raise potatoes in order to eat. There was no state because the possessions of the tribes were so meager that there was nothing of real consequence to redistribute. And while fights between tribes happened all the time, these were brief and sporadic, more like quick raids than anything else. If you will, these peoples had no follow-through.

The state comes about in the following way: As time went by, the hunting tribe began to change. Tribes looked at the herds and flocks of the animals they hunted, and they could recognize certain similarities to their own behavior. After all, being ancient or being tribal is not a synonym for being dumb. It is our different environments, as shaped by our evolution and our impact on our evolution, as well as the way we use our "smarts" today that makes the difference.

The state would eventually develop out of these changes.[21] Let's follow the hunters: They would seek out large game animals that moved about in herds and followed seasonal migration patterns. It was thus possible for hunters to regularly cull the herd, in somewhat the same way as wolves might. And at some point, as the hunters followed a herd, they became less hunters seeking some single animal in the forests or on the savannahs, and more and more, and finally almost exclusively, the caretakers of a herd. It had occurred to the leader of the tribe that the tribe could prosper if it took a particular herd as its own, followed it regularly, perhaps even guiding it, maintaining it at its optimum size, and protecting it from other predators. He communicated this revolutionary notion to his followers and they acted on it. Thus a new pattern developed, tied even more tightly around the seasons. The crude hunter became a herder, with a year-round concern for what was no longer just *a herd* of cattle, reindeer, goats, or whatever, but *the tribal herd.* The tribe began acting less like wolfish predators. For example, rather than just killing any animal in the herd, they spared the females during the calving season. The herd had become tribal, communal property. Protecting the herd, not hunting, became the tribe's primary occupation, and they would defend their herd to the death against other predators, human or otherwise, for the maintenance of the herd in proper numbers and health was truly a matter of survival for the tribe. Our ancestors' immediate rage at others and his "us versus them" mentality was useful.[22]

In this way the tribespeople had become, by stages and degrees, dependent upon the herding way of life. If they lost their herd through disease or misadventure, they might try to hire themselves out to another tribe as sheep, goat, or cattle herders. The more successful tribe might or might not take them. Or these herders without herds could attack another tribe and try to take the others' animals for themselves. Other than those two possibilities, they would die of starvation. Over the centuries, some tribes managed to increase rather substantially while others were absorbed or otherwise died out.

Developing as well were the primitive agriculturalists. Advancing in stages, planting by poking holes in the ground with a stick and then inserting seeds, later by drawing a crude plow by human labor to make a furrow, and even later yet by domesticating animals such as horses or oxen to draw the plow for them, the peasant cultures emerged slowly. Some of these may have been matrilineal, which was almost totally unknown among hunting and herding peoples, and represented a quite different process of development.

In developing the use of animals to pull carts and using wheels (which had originally been mill wheels used for grinding grain) for the carts, they set the stage for their own possible slaughter or, later on in development, their serfdom. Herders and peasants lived near to each other, and the herders saw what the peasants had done. In some places, herders had domesticated their own animals, such as horses, which they found useful to ride about and use to help herd their meat animals. Whichever way it came about, in some places herders hooked animals such as horses to carts, and rather than using the carts to haul grain to the millstone as they had seen the peasants do, they mounted the carts; or they mounted their horses directly, hoisted their weapons and swept down on the peasants, killing them village by village and pillaging the smoking ruins at will. The slow, poorly armed and nearly immobile peasants were no match for a tribe of herders mounted and equipped with the tools, the weapons, of their trade,[23] formerly used mostly for killing wolves, foxes, and other human herding or hunting tribes. After all, there is usually a dispute between tribes for land. The peasants wanted to farm and so tried to keep animals out of their corn or wheat; yet the farmers' land was also prime grazing land. The herders coveted the fields, especially if there had been a drought and their cattle had not fed well. So the herders attacked and gained great wealth. They had been productive in their own way, but they had no surplus. Now they killed the peasants and took their land, seized their grain, cut down their trees and ate their fruit. The herders had discovered a new occupation, that of warriors, and they moved to it readily since it paid off so well for them. As Diamond points out, the warriors needed the peasants to produce this surplus.[24] They hired people from failed or weak herding tribes to herd *their* animals (note how very different this is from the raids on and butchery of early hunting tribes), and seized the goods of the peasants. Being a warrior not only paid better but it was also a lot more exciting to disembowel someone than watch your goats graze all day.

In time, the treatment of the peasants would reasonably change. The warrior had been a manager of a herd of animals, but now used underlings to do that job. The example of those lesser figures from less successful tribes who had become clients of the warriors perhaps gave the warrior chief an idea. He and his fellows were skilled at managing their own herd and their clients, so could they not apply that set of techniques to the peasants? After all, dead peasants plow no fields, grow no trees, brew no beer, supply no food. Suppose, the chief must have mused, he put the peasants to work for him? Obviously some things would be different from the clients, who were herders as he had been and therefore at least marginally human, but the peasants were merely human-shaped animals, possessed of squalid habits and dedicated to grubbing in the soil. They were more like cattle than like his hired hands. But he could control them in much the same manner as he did his horses or cattle, he could make them produce, and he could make them give over their surplus to him at every harvest. Would not the tribe with many peasants to work for them, like the tribe with many herds of cattle, be rich?

Note here that movement—ethical movement, we might say—has taken place. "Us" was no longer just the original tribe. There were other people, who, even if they were only clients, might be people. Life was not so simple anymore. It was not the strict "us versus them"; there were gradations, there were degrees of being human outside the tribe. If peasants were not human, at least some other people were. Indeed, with some tribes, there had been a gathering of many different tribes together, and as a clear precursor of the feudal state, in these megatribes (the Mongol hordes, for example), one finds that even more recognition of tribal likenesses had penetrated and acceptance of humanness had been achieved.

So now the warriors would sweep down on the peasants at harvest time. The peasants would flee in terror, but the warriors would round them up, killing as few as possible in the process, and give them the word, the new deal, so to speak. The peasants would turn over only their surplus to the warriors, and they would be spared.[25] The warriors would leave the huts standing, the trees unburned. Of course they would play a few boyish pranks, a few women would be raped, but then it is very hard to change old habits quickly.

So much for the conquest. The warriors provided no services, save that of keeping other warrior bands from doing as they were doing. The warriors grew no grain, tended no crops, they did not even tend their own flocks and herds anymore—others did all the productive work. All the warriors did was to promise to look upon the peasants as labor motors, rather than as low and dirty animals in need of extermination. Providing nothing, the warriors were exploiters, maintaining their position by force and the threat of force. Even so, the new deal was an evident improvement for the peasants. Far better to be exploited than killed.

Does this sound cruel? Certainly it will to many who will read this, and it sounds that way to me. But does it sound that way to those who today are engaging in indiscriminate slaughter in the Middle East or Darfur? Hardly. It is one's way of life. As Garrett Hardin points out, until quite recently, "The *idea of cruelty*—i.e., cruelty as something to be abhorred rather than enjoyed—scarcely existed."[26] And there are still many tribal people left in this world who would enjoy hacking off someone else's head or bombing someone else's children. It may not be necessary to go as far as the Middle East to find them.

Here we have the beginnings of the tribal state, the largest and most advanced of them arising from the megatribes.[27] The warrior tribe, in seizing a number of peasant villages to loot and terrorize, acquired for itself a kind of form, a territorial expression or definition. The warrior chieftain would look out from his tent and proudly recognize that *this* village was *his*, even as he also recognized, quite reluctantly, that the village across the river belonged to the leader of another band, or at least would until he attacked next spring. Owning humans of other tribes is certainly one of the earliest forms of property, just as the conquest of them had been an early contact sport.

From being entirely nomadic, his warrior tribe began to move among the various villages and became more domesticated in the process. In settling down, the warriors had started to lose one of the points—their mobility—that differentiated them from the peasants. Moreover, out of the crudest sort of self-interest, the warrior had been led to accept the value of others not like his tribe. Other people were needed to do the dirty work he would not do, to provide for his needs, to supply his wants. The warrior depended on his peasants, however loath he must have been to admit it. If this limited notion of interdependence is regarded as no great ethical theory today, it must be put in the context of its time and recognized for the great advance it then was.

His villages! How the warrior chief must have regarded them—and himself—with pride, so many people working on his orders, out of fear of his power and majesty. But always there was a threat from across the river or from beyond the hills. Other tribes and other chieftains existed with their villages, always quite willing to make a raid onto another's territory and engage in those oldest of human games: burning, raping, and looting. Always the warrior must stand guard lest his village suffer attack and have no surplus left to surrender to him. This is the age-old responsibility of the warrior, to protect the lives and goods of those whom he exploits from other exploiters. It might even have been that the warrior would take the opportunity to point out to their villagers that they lived only under the mighty protection of their warriors, who would ensure that the other warrior tribes would not completely destroy them.[28]

If this responsibility to take care of one's own property, one's serfs, arose out of self-interest, it was nonetheless binding on those same self-interested grounds. While it was not thought that the peasant had a right to life, the notion that he had some sort of limited right to be protected against the violence of outsiders had gained acceptance.

Over time the idea of tribute, which benefited the serfs, came into being. No longer would the warrior horde get the chance to come rampaging into the village, armed to the teeth, ready to fight when necessary, or just to get drunk and bash in a few skulls for the fun of it. Instead, the peasants, in their sly acceptance of docility and to avoid such a confrontation, quite literally beat the warriors to the punch and went to the brawlers' encampment the day before. Usually the warriors would stop just over the hill or around the bend before riding into the village in the morning, and so the peasants would forestall them. They would bring the surplus to the encampment the day before it was due, with a lot of bowing and scraping on their part, done to make the warriors understand how brave and fearsome the warriors were, how the peasants feared and respected them. The peasants would then persuade the ruffians to scoop up their loot and leave. I suppose the warriors might be halfway home before it dawned on them that they hadn't had their usual bits of fun with the peasant women.

So the benefits to the peasants were obvious, and if the warriors still itched for some action, a sneak attack against the villages of someone else would easily make up for it and provide them with joy and merriment.

Season followed season with monotonous regularity, and the warrior band fell into the routine of its relationship with the cultivators. Once entirely nomadic, bounds had been set to the wanderings of the warrior tribes; political interests had come into play that restricted their conduct more and more. The relationships between conqueror and conquered became more expanded and complex. Ultimately the wandering ceased (occasional raids and wars to the contrary) and the warriors began to settle down in their territory. They settled first as a group, but as the pressures of battle and violent plundering slackened, they began to move apart from each other, first in camps to be moved at the first sign of danger, and then in more permanent dwellings, perhaps even in the first rude castles, constructed, of course, by peasant labor and a special levy. These changes were driven by the needs of the time. The warrior chief would send his fighters out into the territory, one family perhaps to their new dwelling on the hill overlooking the river, with the enemy on the other side and the little farming village on their own side. Another warrior family would go to the hill near the pass through which a potential invasion might move to attack yet another village, and so on. The moves were dictated by strategic reasons. So from wanderers in *the* land, the warriors had become settlers in *their* land, settlers now living farther apart from each other and closer to the peasants they controlled.

So developed slowly, blindly, haltingly, the process Oppenheimer calls "amalgamation,"[29] the coming together of warrior and peasant tribes to form what eventually would be one state, with a very strict class structure. Probably the first element of amalgamation was language. Peasants needed some way of indicating their surrender and submission, and the warriors needed a method of indicating the goods and services they required. The result would be not just a mixture of the two dialects or languages, but a mixture reflecting the appropriate dominance and subservience of each party. For example, in the present-day English language there is the word *beef*, which is meat from a cow, cooked or otherwise prepared for eating. And since beef is what is put on the table and is a top-quality, high-protein food, it comes as no surprise to find that word is derived from the language that the Norman conquerors brought to England. The related term *beefeater* refers to the British Yeoman Guards, who originally were paid in beef. On the other hand, the word *cow* refers to the animal on the hoof, needing to be herded and otherwise tended. Since the Saxons, being conquered, comprised the lower class who would tend the animals but not be allowed to eat them, this word comes from the Saxon language.[30] These two words seem to be quite unrelated terms in our language. They refer to the same thing, but they do so from the standpoint of different tribal peoples.

Religion, especially in Europe, also was a fit way for drawing a new nation-state together and blending the different stocks. Certainly the common practice was for the religion of the conquerors to be dominant, with their chief deity retaining the superior position. The god or gods of the conquered become subordinate figures in the pantheon, or perhaps as happened in Europe, the gods of the peasants became the devil of the conqueror's faith.[31] Witches, the stubborn defenders of the old peasant religion, were sought for torture and burning, because at that time, to offend by refusing to accept and worship the god of the king was to commit both treason and heresy at the same time.

Thus throughout society, in all the pathways of myth and social custom, amalgamation occurred as different tribes gradually became one society. Strangers were those who lived across the river or in the next valley, who had different languages, gods, or kings. Tribes became higher and lower classes of the same society as they became more familiar with each other and established similar although not identical customs, habits, and traditions. A more unified way of life developed. This is true wherever one looks in the world, whichever continent or people or time is studied. In tiny Switzerland, for example, the first people (or at least the first we know of, the Lake Dwellers) probably came from Asia and settled perhaps as early as 4000 BCE. They probably herded cattle. They were attacked and exterminated by Celtic tribes who were well entrenched by around 400 BCE. One of the Celtic tribes, the Helvetii, was conquered by the Romans and used as a buffer against the wild Germanic tribes to the north. When Roman power in the region faded, the Helvetii were conquered by the Alamanni and the Burgundians, although yet another German tribe, the Franks (who came to be controlled by the Habsburgs and turned into a very large feudal state), overcame them all. Switzerland did not win independence until 1291, when three small tribes declared their independence, fought the Habsburgs, and won by using guerrilla tactics. They established a league, or constitution, in which they mutually promised to aid each other. These peoples were, according to their constitution, the "people of the valley of Uri, the democracy of the valley of Schwiz, and the community of the mountaineers of the Lower Valley."[32] Rather than the usual contest of tribe against tribe, the Swiss overcame that parochialism out of a shared interest in independence from the people beyond their little region—the foreigners—and developed a sense of collaboration between equals that helped the Swiss grow and become strong in their region. Within each canton or tribal grouping, however, old tribal and class distinctions would return to plague them, and would ultimately result in their being conquered by Napoleon.

On the other hand, there is giant China. The Shang (or Yin) dynasty ruled some areas as early as 1700 BCE, and even then their success was over an earlier dynasty. The Zhous followed and seem to have been quite decentralized, based rather strictly on tribal and familial bonds. Of course the dynasty one might think of first was the Ch'in, which followed a different path in taking over, and centralized most of China quite thoroughly. The Han dynasty followed and when it collapsed, the warlords, or local tribal leaders, took over. The political history of China is the story of a central monarchy building a dynasty, trying to centralize power, and being torn apart by the local regions and tribes—the Han, the Dai, the Lahi, the Yao, the Qiang, and many others. All the typical fluctuations of tribal politics, mingled with attempts by ambitious leaders to create great monarchies for themselves, make up much of the history of China, at least the part of history that the political historians tell.[33] Take any country toward the end of its tribal period and the beginning of feudalism, and you find a very rough similarity with these processes commonly occurring.

A key development in this later stage of tribalism leading to classes and statehood was the development of a sense of justice, with the ruling class incurring certain obligations of a rude, rough and ready sort.[34] This justice arose essentially out of an ambition on the part of the rulers to exploit their inferiors in a more efficient way, to manage their slaves, their human herd better. Of course in the beginning it was felt sufficient to simply take the peasants' surplus and leave, protecting them in case of foreign attack and no more, but along the way it occurred to the rulers that a closer supervision could provide a higher yield. This stage of development begins occurring as a people starts making the change from pure tribalism to a feudal state.

Before we move along, however, let me stay with tribalism for a bit and point out a few more items. After all, it is the first stage of human development, the stage in which we remained the longest. It is a stage in which many people remain today, a stage that still influences the conduct of many people as one can see in the present turmoil and war in the Middle East.

But the concept of tribe is not obvious to most people in the United States or some Western European states. We have little tribal sense here. Ask any crowd of people in this country which tribe they belong to, and you will mostly get blank stares. On (or close to) an Indian reservation, people will know what you are asking, but probably nowhere else. But I was surprised, in talking with my wife about this, how prevalent tribalism is in many places in this country. In the small Utah town where she went to school in the late sixties, there was a distinct four-class system, with everyone in the school sitting with and talking to only their own people. The Utes and the Navahos sat in back, and thoroughly disliked each other; toward the front, there were the non-Mormon whites and then at the very front the privileged Mormons, who regarded everyone else as inferior, led in this belief by a Mormon teacher. In the 1970s and 1980s in Oregon, she saw the same kind of tribal division in all aspects of life: it was the Klamath Indians versus the Modoc Indians, who often hated each other and lived as separately as possible—one does not collaborate with the enemy, and if you are not a member of *the* tribe, you are an enemy. Indian organizations in this country often seem to want to foster tribalism and keep "their people" back on the reservation, still living the tribal culture of the fifteenth century. Change—and therefore amalgamation—is bitterly opposed.[35] It may be too late for that, as many (and in some tribes, most) of their people have in fact amalgamated.

In the most important sense, it is good that we are advanced enough so that belonging to a tribe is not significant to most of us. It is unfortunate in two ways, however: one is that we don't understand a great deal of what happens elsewhere. In the stories about Iraq, after the mess of the invasion, what stands out for me is that US troops in that country, newspaper reporters sending back stories on the country, and dozens of presumed experts and talking heads generally failed to think about or discuss the underlying tribal basis of the Iraqi society and state. Saddam Hussein knew it well and exploited it efficiently and brutally, whereas we seemed to think of conflict there as just the conflict of religious sects. And without understanding what the Iraqi state was like, the US government's efforts were bumbling and uninformed.[36]

The second point I would make is that when a tribalist reaction occurs in our own country, we rarely understand ourselves. During the great dispute over illegal immigration starting in 2005, a great many Americans quickly fell into an "us versus them" attitude, in which facts were disregarded and these immigrant people, mostly hardworking and decent folk, were attacked as an enemy. The harder they worked, it seemed, the more many people hated them; the more the immigrants tried to live the old and apparently outdated American dream—which said if you worked an honest day's work and saved your money, you were a good person—the more people turned on them, and "minutemen" and border patrol personnel hunted them like dogs. I was listening to one of those awful radio shows where ordinary folk are begged to voice their opinions, and one woman said that people who came here to work were agitators and brought nasty diseases. Thereafter, I would think of this belief as the *cootie theory* of immigration. And the day I was writing this, ABC News reported on a high school senior, ready to go to college, who was arrested, handcuffed, and hauled off to prison. This young man, brought to this country by his stepfather from Germany, was never legally adopted and hence technically an illegal alien.[37] When my online service ran one of those totally unscientific polls asking people their opinion about these circumstances, almost one-fifth of the responders indicated that they were fine with it—arrest the little wretch, throw him in jail, and then boot him out of the country. I wonder how many of that 20 percent would rather have stoned him. Tribalism is not dead anywhere.

What is happening here is that we are demonstrating that many of us are not as advanced as we ought to be in order to function in the modern world. The old tribal reflex, "outsider enemy, tribesperson friend," while no doubt appropriate for us when tribal living was all that was possible, is counterproductive today. It is good to note that 80 percent of the people did *not* have that attitude. They moved beyond it to a more productive and appropriate approach. Of course, economists can and do argue over various points regarding immigration, and just how admitting this or that class of immigrant, or how many should be admitted, can affect the economy of the nation. But economics is not the question and never has been in this country or any other, whether one goes back to the days when anyone who could get here got in, or whether one argues through all the laws and schemes of the present. Nor is it a question of dealing with the "melting pot" theory and pretending that this or that ethnic group "refuses" to assimilate. Recently someone translated "The Star Spangled Banner" into Spanish, to the horror of the nativists and opponents of any kind of cross-cultural "contamination." This translation is bruited about as some kind of example of Hispanic intransigence, or worse, as an attempt to take over the country.[38] Who remembers, as Roger Lowenstein points out, "ninety years ago, some signs were in German, as were 500 newspapers on American soil"?[39] Or that the then United States Bureau of Education, in 1919, had "La Bandera de las Estrellas" done up in what was apparently the first translation of the national anthem into Spanish?[40] As economist David Card has said, the "economic arguments . . . are almost irrelevant."[41] It is, alas, the tribalist cootie theory that is the deciding one for many people.

Other instances of tribalism in our time can be pointed out in various ways. We could talk of patriotism as calling up the "us versus them" sentiment, as suggested earlier. But I suppose the simplest is to mention the recent report from Colombia about the primitive folk who just emerged from the jungle. Many South American countries have tribes: Brazil, for example, may have as many as two hundred of them, most quite small. These people usually speak only their own language, so one may have as few as four or five hundred people speaking their own unique language. But the report I have in mind comes from May 2006, when around eighty or so members of the Nukak-Maku suddenly popped out of the Colombian jungle, willing to give up their tribal life for the benefits of modernity. They were nomads, wearing few clothes and living off a diet of boiled monkey and berries. "We do not want to go back,"[42] said a member of the tribe. As perhaps the most evident example of how tribal life differs from modernity, "When asked if the Nukak were concerned about the future," the spokesman did not understand. "'The future,' he said, 'what's that?'"[43]

If one would like contemporary examples of tribal life elsewhere, one could go to the Northwest Frontier Province and surrounding area of Pakistan, where tribes such as the Mahsuds, the Afridi, the Wazir, the Bhittanis, and others, mostly Shiite Muslim, wage continual wars and skirmishes against each other. This is the area where many people believe Osama bin Laden fled when he left Afghanistan. It is a harsh and inhospitable land, nominally part of Pakistan, but the authority of that state is limited and often nonexistent in the region. It is and always has been a battleground of tribes, whose way of life is almost unchanged over many generations.

Or if one wishes to talk about tribes in Europe, journey to the territory of the former state of Yugoslavia. Here for many years Marshall Tito, the semi-communist dictator, held sway. He ruled his people with a heavy hand, making sure that his subordinates were loyal to him over and above any tribal loyalties. When he died in 1980, it took only eleven years for the country to fall apart. The Serbs, the Croats, and the Slovenes all went their own ways. Now we have such tribal states as Bosnia and Serbia to reflect this. It may be seen in this case that while a powerful totalitarian ruler may force tribal peoples to cooperate, this works only briefly—something we are now seeing as well in Iraq. Force does not help people to transcend their tribalism; it just covers over the divisions for a while. It might be interesting sometime to compare the rule of Tito of Yugoslavia with that of Saddam Hussein. Both ruled a diverse state of tribal peoples with an iron hand, and when their rule was ended—one by death, the other by invasion and capture—their respective states disintegrated in tribal-religious warfare. Neither dictator succeeded by force in bringing people together for much longer than the force lasted. These rulers retarded the growth of their people, who have had to start the development process over again to rise beyond it.

Tribalism as a first stage of humankind is apparently also the most enduring. If people are quite fitted for that stage and only that stage, it seems harmful to force them out of it. One must wait, as one waited for the Nukak.

The use of tribes even today has great value in certain areas, as the case of Iraq points out. Overwhelmingly tribal peoples, the Iraqis have seen many tribes destroyed by war as well as by the brutalities of the government before the war. Typically, this kind of tribelessness happens when an area is denied normal growth because it is conquered repeatedly by colonialist powers, which artificially divide it up into administrative regions with no regard to tribal boundaries. This has happened in the Middle East just as much as it has happened in South America and Africa. Iraq was first established in relatively modern times by the Ottoman Empire, then taken over by Britain, then it became nominally independent but was in fact under US control via its ally, Saddam Hussein, who was used to try to control Iran.[44] Now, with the dictator out and the United States allegedly trying to establish democracy in a country that does not want it because it is unready for it, tribes may reemerge to play a significant role in controlling the warlords who claim a fundamentalist Islam as their justification for brutality. Of course tribes and warlords alike will be brutal, for the latter are acting outside all human control and human regard and the tribal leaders are trying to regain control. When you take a people who are only able to operate efficiently in a tribe and force them out of the tribe, they frequently become extremely violent. If tribal violence is cruel, this tribeless violence is surely worse as these people will attack anyone, even someone who was in their tribe when their tribe existed as a functioning group, so any restoration of tribal control would be not only beneficial, but a necessary prelude to later development.[45]

John Stuart Mill points this out to us early in his study on what we would today call "democracy" when he mentions that "people whose passions may be too violent or their personal pride too exacting, to forego private conflict and leave to the laws the avenging of their . . . wrongs . . . are not capable of advancement."[46] To suppose such people capable of immediate political progress is wrong, and if you deprive them of their tribes, the one great constant in their lives, they become disoriented and are capable of any atrocity. Without the tribe, tribal people are left naked and disoriented before the world. They find in hatred the unifying and motivating force that draws them beyond their own pitifully insecure selves into, as Eric Hoffer says, a "flaming mass" that destroys all in its path.[47] And so it is that men who need tribes, that is, need the male bonding and can no longer have it in a tribe, will unite under any leader—Charles Taylor of Liberia comes to mind, or in Iraq, al-Sadr—who will lead his males in the cruelest ways of torture and destruction possible.

Fortunately for people there, tribes have reasserted themselves in parts of Iraq, especially the rural areas, where some tribal power always remained under Hussein, unlike in larger citylike areas. The new, American-sponsored central government is an absurd joke. It has no power outside the "green zone" of the United States, and not much influence there. The outside troops that remain are subject to attack from all sides just like the natives. They cannot maintain order. But the tribes can, as in the town of Amara, capital of the province of

Maysan. Here we have "a hopeful tale of small-town camaraderie, fierce independence and, above all, tribal power."[48] What happened? After serious violence began, one tribe invited all the area sheiks, or tribal leaders, to a meeting. These leaders agreed that they would jointly punish those who attacked any tribe or tribal member, and that they would carefully watch and supervise all militia members. Since then, the violence there has dropped off almost to the vanishing point. Music shops are now open and the man who runs the domino parlor and coffee shop is not faced with armed thugs coming by and forcing him to shut down, claiming that playing dominos is not Islamic. Tribes work where men believe in them and they are allowed to function. When people are ready, they will evolve beyond them. It is counterproductive to try to force people out of tribalism. And it is preposterous to try to force these people to be democratic or to adopt the humane standards of people at later stages of social development. Humane conduct and advanced behavior are not created out of whole cloth, nor are these kinds of behavior enforceable at gunpoint. These values are only acquired peacefully and when one is ready for them. Genetic development as well as interest, economic or political, must be there first.

Natural progress and development, where people learn for themselves what is better, works slowly and is the only efficient process we know for development. After all, we do evolve remarkably rapidly.[49] A wider distribution of power as found in democratic states can be a result of such changes and a catalyst for more changes. Changes can also be discovered and developed in such simple yet fundamental matters as giving women entrance to the military, increasing development of female groups and alliances for mutual support,[50] and institutionalizing political power, rather than resting it in leaders. We need not wait until democracy for all these things, as people start to grow and change long before democracy becomes possible. We can see it in the feudal and especially late feudal states, which can and have eliminated many of the most outdated, violent aspects of tribalism. If violence is no longer a successful evolutionary strategy, then a wider cooperation in economics and other areas is needed. This becomes more clearly envisioned as much more useful to the individuals involved, and since in fact it is so, we will begin to evolve in less aggressive, if not completely nonviolent directions. And it appears that is just where our changing evolutionary development has gone. Just how it has gotten there, and how this development works itself out in the increasingly state-organized rather than tribally organized worlds of feudalism, will be seen as we move further along our path.

NOTES

1. William Tenn, *Of Men and Monsters* (New York: Mentor Books, 1968), p. 11. The narrator in this fine science fiction story is appropriately speaking of the men in his tribe.

2. Richard Wrangham and Dale Peterson, *Demonic Males: Apes and the Origins of Human Violence* (London: Bloomsbury, 1997), p. 25.

3. This is the report of the ancient historian Herodotus. It has been confirmed more recently by anthropologists. http://www.silk-road/art/sarmatian.shtml (accessed July 12, 2006).

4. Wrangham and Peterson, *Demonic Males*, pp. 64–72.

5. Ibid., p. 79.

6. Ibid., p. 173.

7. Kate Connolly, "Hungarian Descendants," http://www.telegraph.co.uk/main/news/2005/01/12 (accessed July 12, 2006).

8. Wrangham and Peterson, *Demonic Males*, p. 177.

9. Randolph Bourne, "War Is the Health of the State," an essay he was working on at the time of his death. See it at http://www.bigeye.com/warstate/htm, p. 4 (accessed September 3, 2006).

10. Ibid., p. 6.

11. Plato, "The Statesman," trans. Benjamin Jowett, in *The Dialogues of Plato* (New York: Random House, 1920), 2: 294.

12. James Lipton, *An Exaltation of Larks, or the Venereal Game* (New York: Grossman, 1968). It is a study derived from the ancient practice of venery, which tries to give appropriate names to groups of various sorts, as in a "plague of locusts," or more amusingly, "a smirk of courtiers," or a "blast of hunters."

13. Wrangham and Peterson, *Demonic Males*.

14. David Riesman, Nathan Glazer, and Reuel Denney, *The Lonely Crowd* (Garden City, NY: Doubleday, 1953), p. 28.

15. Ibid., p. 33.

16. James Bowman, *Honor: A History* (New York: Encounter Books, 2006), p. 4.

17. Ibid., pp. 320–21.

18. Ibid., pp. 1–2.

19. Franz Oppenheimer, *The State,* translated by John Gitterman, introduced by C. Hamilton, http://www.franz-oppenheimer.de, pp. 11–32 (accessed June 14, 2007).

20. Jared Diamond, *Guns, Germs, and Steel* (New York: Norton, 2005), pp. 268–69.

21. Oppenheimer, *The State*; see also Diamond, *Guns, Germs, and Steel,* pp. 282–92.

22. Wrangham and Peterson, *Demonic Males,* pp. 74–77. See also David Berreby, *Us and Them: Understanding Your Tribal Mind* (New York: Little, Brown and Company, 2005), p.16. Note the difficulty Berreby has in not accepting that we are capable of growing out of our "tribal mind."

23. Besides Oppenheimer, *The State,* see Diamond, *Guns, Germs, and Steel,* pp. 90–91 on the role of horses, etc., in agricultural pursuits and warrior efforts. Robert Ardrey, *African Genesis* (New York: Dell, 1967) was perhaps the first to point out that when anthropologists talked about primitive people using "tools," what they meant but failed to say out of their generation's ideals of political correctness was that calling weapons "tools" was just covering over the hard truth that tribes killed each other, and that the chief tools of herding

tribes were the weapons they used to kill other peoples. See page 207 and following. Of course there were occasions when states were not formed by herder-warriors but by agriculturalists. When herding was no longer possible, perhaps because of climate change, then warriors arose out of the most aggressive and largest farming tribe. The Olmecs of Mexico, an early megatribe, are an example. It appears that warriors are a necessary human development at this point: these farmer-warriors were just as ferocious and cruel as any herding tribe. For a study of the Olmecs, see Amber M. VanDerwarker, *Farming, Hunting and Fishing in the Olmec World* (Austin: University of Texas Press, 2006).

24. Diamond, *Guns, Germs, and Steel*, p. 406.

25. Ibid., pp. 90–91, and, of course, Oppenheimer, *The State*.

26. Garrett Hardin, *Nature and Man's Fate* (New York: Rinehart and Company, 1950), p. 303.

27. See the discussion in Diamond, *Guns, Germs, and Steel*, about the early expansion of these, which he calls "chiefdoms," especially pp. 270–76.

28. Diamond, *Guns, Germs, and Steel*, p. 277.

29. Oppenheimer, *The State*, pp. 34–35. Note as well on 35 his comments on "solidarity," a sense of people of different tribal origins now belonging together. Compare to Diamond, *Guns, Germs, and Steel*, pp. 276–77, and how leaders in the state-formation business, called "exploiters" or "robbers" by Oppenheimer, are the "kleptocrats" of Diamond.

30. Any large dictionary can provide one with the word origins. And http://www.toweroflondontour/yeoman.html is interesting and informative.

31. This is from the viewpoint of the rulers involved. For interesting and perhaps tongue-in-cheek speculation about how old gods might feel about being so summarily dismissed, see the novel by Manly Wade Wellman, *The Old Gods Waken* (New York: Doubleday, 1972), especially pp. 68–69.

32. The constitution is quoted in W. D. McCracken, *The Rise of the Swiss Republic* (New York: Henry Holt, 1908), pp. 87–88. A good deal of Swiss history can be found on the Web. For example, http://

history-switzlerland.geschichti-schweiz.ch/. Another site with some interesting maps showing both the ancient configuration and present cantons is http://www.rellint.com/roll/switzerland.htm (both accessed September 2, 2006).

33. There are an incredible number of Chinese histories. Online, one might wish to consult, among others, http://www.c-c-c.org/chineseculture/minority/minority.html, or http://www.china.org.cn/e-gudai/index-1.htm (both accessed September 2, 2006). Some brief but insightful comments from a political scientist can be found in K. J. Holsti, *International Politics*, 7th ed. (Englewood Cliffs, NJ: Prentice Hall, 1995), pp. 25–35.

34. Besides the discussion in Oppenheimer, *The State*, see Diamond, *Guns, Germs, and Steel*, pp. 276–78.

35. See http://www.aimovement.org/ggc/history.html. For example, "Desegregation was not a goal (of the movement). Individual rights were not placed ahead of the preservation of Native Nation sovereignty." In short, defending the rights of individuals is not the goal. Preserving the powers of tribal leaders over their members is. See also http://www.americanindianmovement.org/papers/struggle.html. A recent example is the vote taken by the Cherokee Nation to exclude people who had belonged for generations. The exclusion was because those people were black, and black is not a defining characteristic of that tribe. See the article by Tim Reid, "Cherokees Disown Slave Descendants," in the *Australian*, March 6, 2007. It can be found at http://www.theaustralian.news.com.au/story/0,20867,21331738-2703,00.html. (All accessed March 10, 2007.)

36. Some of this has changed. A good place to start looking at this is Amatzia Baram, "The Iraqi Tribes and the Post-Saddam System," Iraq Memo #18, July 8, 2003, posted at http://www.brookings.edu. This and similar studies and reports are now—finally—being given credence (accessed September 7, 2006).

37. Jessica Golden, "After Graduation, Teen Faces Deportation," *ABC News*, May 21, 2006.

38. For example, "Bush: National Anthem Should Be Sung in English," Associated Press, April 28, 2006; or for a more general

view, David Montgomery, "An Anthem's Discordant Notes," *Washington Post*, April 28, 2006, p. A1.

39. Roger Lowenstein, "The Immigration Equation," *New York Times Magazine*, July 9, 2006, p. 70.

40. Http://memory.loc.gov/cocoon/ihas/loc.natlib.ihas.10000 0007/pageturner.html (accessed September 17, 2006).

41. As quoted in Lowenstein, "The Immigration Equation," p. 43.

42. Juan Forero, "Leaving the Wild, and Rather Liking the Change," *New York Times*, May 11, 2006, p. A1. On dietary grounds alone, we can sympathize.

43. Ibid.

44. See as a brief summary, Roger Cohen, "Iraq's Biggest Failing: There Is No Iraq," *New York Times*, December 12, 2006, p. Wk 1.

45. Anthony Shadid, "In a Land without Order, Punishment Is Power," *Washington Post*, October 22, 2006, p. A1.

46. John Stuart Mill, *Considerations on Representative Government*, ed. Currin V. Shields (New York: Library of Liberal Arts, 1958), p. 7. Indeed the entire first chapter is very useful.

47. Eric Hoffer, *The True Believer* (New York: Mentor, 1962), pp. 85–86.

48. Sabrina Tavernise and Qais Mizher, "In Iraq's Mayhem, Town Finds Calm through Its Tribal Links," *New York Times*, July 10, 2006, p. A1.

49. Wrangham and Peterson, *Demonic Males*, p. 43. See also Patrick D. Evans et al., "Microcephalin, a Gene Regulating Brain Size, Continues to Evolve Adaptively in Humans," *Science* 309, no. 5741 (September 9, 2005): 1717–20, and many other recent sources.

50. Wrangham and Peterson, *Demonic Males*, p. 243.

CHAPTER 2

FEUDAL PEOPLE

Nothing on earth has the same power of moving a man to violent and implacable hatred as a member of his own species.[1]
G. Molinari

The tribal stage of development, provided a people managed to survive and partly surmount it, turned into feudalism, which is a stage of development that can occur anywhere, not just in Europe. It was a slow process when it developed, and for many peoples today it still has not happened or is happening now. It grew out of tribalism under certain conditions—a number of primitive tribal states clashed with each other and feudalism arose out of those brutal little wars. So far, we know that the leaders of hunting and then herding tribes wound up controlling peasant villages and having little conflicts and battles with surrounding tribes doing much the same thing. It remains to point out that the development of feudalism out of tribalism is this ongoing process. It is difficult to figure exactly when tribalism becomes less dominant than the feudal arrangements, because tribalist sentiments remain with people as a background to their new obligations and allegiances. Let me show this by picking up the story with a discussion of justice, which is an internal, developing obligation of upper class to lower class, and going on from there. It is not a matter of obligations between tribes, because that kind of obligation does not exist, or exists quite rarely and precariously.

The chieftain led his tribe into battle with other tribes. If the chief won, he accumulated more territory. Territory is useless without people to work the land, so inevitably the chief wanted more serfs, who would be the tribal peoples he had conquered, the "labor motors"[2] he would use for his benefit. No longer was extermination a goal. Even small children were allowed to live. After all, they would shortly grow up and be able to work. But there was a natural limit to conquest. A chief could expand his domains only so far before he found he needed subordinates and, in effect, appointed them his generals and sent them forth to conquer. When they succeeded, he rewarded them by giving them the new land to operate for him. These generals were barons now. Included with the land were the serfs to work it. Land and serfs were the wealth of the feudal state rather than the herds of the tribes. The chief had control of enough wealth that, rather than being a part-time leader, he was a full-time controller, a politician, a king. And this king had appointed his generals as his nobles, above and beyond their role as barons, so they became the members of his court. They were obligated to support him, and the more deeply we get into feudalism, the more explicit the promises were between the king and his subordinates or vassals. So many days of combat service each year, so many knights, so many foot soldiers, and so on were promised to the king in exchange for one's becoming a lord; and in time a lordship or other title of nobility, originally granted for success in battle, became the hereditary possession of a noble family, subject always to the promises of support, the feudal obligations. As McIlwain says, once a feudal agreement was made, the grantor of the land and serfs, the king, would be the lord of the baron, "with definite rights over him and definite obligations to him of a continuing character, and he as vassal had reciprocal duties and rights in relation to the lord."[3]

Suppose you were such a lord; and let us suppose that as Baron/General Pushgas, you won land for your monarch, of which manor you were then master. Your wealth was derived from this land and the peasants who worked it. If they did not produce wheat and bake it, you would not have bread. It was an unpleasant fact of life, one you generally preferred not to face, that you were dependent upon what they produced. There was no money economy; all you got was what was done on the land of your manor for you by your peasants. Your wealth was therefore based entirely upon your land and your peasants, nothing more.

But still the thought occurred to you that perhaps your peasants could produce a bit more. If you supervised your peasants more closely, if you managed them more efficiently, your take would be greater. At this point, thinking on the subject was still quite elementary. But even so, you acted. Two of your peasants quarreled about which of them rightfully possessed some chickens. Words were exchanged, and then blows. One was injured. An injured peasant was a nonproducing peasant. Would it not have been more useful for you if you had prevented their quarrel and the subsequent damage? So you stepped in and decided this question. In time, such decisions were made on a regular basis. You informed the village elders that you would come to their village once a month to decide cases. You had them construct a proper chair for you to sit in and you ordered it covered to protect you from the elements. The first courthouse was built. You sat. The first case was between two of your peasants. You told one, "you speak—tell me what your position is," and when he was done, you ordered the other to do the same. It might have happened that the second peasant had a speech impediment that prevented him from adequately presenting his argument, and he asked another member of his family to speak for him. The first lawyer was created. You listened to both sides and then made your ruling. How did you rule? On the preponderance of the evidence. You had no favorites here—a peasant is a peasant. It would have made sense therefore, having no personal stake in who won by your decision, to make a fair decision. If the loser seized the chickens anyway, you would have enforced your decision (a step crucial to your authority) by sending your band of soldiers to beat the fellow and get the chickens back. The first police was created. If you ruled according to common sense, the community of peasants would find it easier to accept and they would obey all your decisions more promptly. Of course there were always a few hotheaded young men who wanted to determine things by the old way of tribal violence, but they were either quickly overruled by the village elders, the former chiefs, or they would be killed. In a key way, the practice of tribal violence had been limited.

When your noble—that is, victorious—ancestors herded goats, they made decisions or judgments on the basis of what benefited the herd, not for the sake of the herd but for the benefit of their tribe. Could you do less for yourself? Notice here, by the way, that as a noble over peasants, you are starting to think less about your tribe and more about yourself. Make a judgment to benefit your peasants and you would have made them more obedient for your own benefit.[4] Justice, even in this limited sense, established a mutually beneficial relationship between feudal masters and serfs. This was the great advantage the feudal state possessed over tribalism, that there was at least the beginning of a judicial system. Crime needed to be punished and it was better for a court to do it, even if the court was only minimally just, than for the punishment to have been imposed by the victim's relatives, whose lack of system "often leads to blood feuds,"[5] which would benefit no one. A better system had been put into place. It was the harbinger of greater cooperation to come, a cooperation first imposed by force and law. It would eventually mellow into something richer, more complex, and much later on in its development, less coercive.

Suppose the dispute was a bit larger. Perhaps it was an argument between two villages disputing over which peasants had the right to farm the field by the oak trees or which group had a right to a well that sat between the villages. Again, you were called upon to decide. How did you decide this? You heard both sides and announced your decision would be handed down in several days. Already you had done well. Your peasants understood you were carefully considering their problem; you, their master, were interested in their affairs. You had indicated the gravity of your concern (and understand: you probably did not care at all, you were just baffled, and besides, you were hung over the day the case was argued . . . but the impression you left was that you cared). The appearance of concern made your serfs more disposed to abide by your decision, and this was what you needed to maintain your position.[6] Back you went to the castle, where you mentioned your bewilderment to one of your knights. He responded that he heard a few years ago about another one of the barons who had a similar case, and how he settled it. This knight was now your legal clerk, and when you rendered your decision, you started out, "In accord with the custom of the kingdom and previous rulings of barons, including the ruling last year." You would have been very impressive. Even the losing side was satisfied; they could say, "Oh, we didn't know that," and obey much more easily than they would have obeyed a command that had only force behind it. This was far better for all than a war between your villages and the consequent destruction of some crops and some peasants. Your interest would have rested in keeping the serfs working as hard as possible to create as much surplus as they could. Regular everyday violence, on their part or on yours, would not have been conducive to such production. Yes, it would have been occasionally necessary to push a malefactor into line, to impress the commoners with the fact that "the law" was backed by force when required, but violence worked most efficiently when it was occasional and when it was clearly called for, that is, when the noble's position could not easily be maintained without its use. Did your peasants care? Not as long as the violence was not so great as to be unbearable, as long as they got the smallest amount of satisfaction. Their lot did not change. They were ruled and they would always be ruled, and besides, they followed tradition: they thought of themselves as tied to a certain piece of land and as the property of their lord, not as individuals any more than the members of tribes had. And they certainly had no idea that they could change anything or that their children would ever be able to change anything. Everything had always been the same, so it would remain the same forever,[7] or so it seemed to them.

A key test of the developing court system and the use of law rather than pure force occurred when we would see the beginnings of the serious protection of peasant rights against a baron; this was the crucial test of the abilities of the kings and nobles to make the system work efficiently. What happens in the state when a member of the nobility committed an offense? What, if you will, would occur when as a baron you found that one of your fellow barons had become a Vlad the Impaler? What happened when this baron went far overboard in his treatment of his peasants, perhaps flogging or impaling many of them without rhyme or reason? At this point the nobility, as a class, were confronted with a serious problem that went directly to the question of their abilities as managers of their human herds. When a noble class could, as a rule, solve this problem, they would probably extend their tenure by centuries and would reap the benefits of their good practical sense many times over. On the other hand, when they could not or would not discipline a member of their own class to preserve the benefits of rule for the entire class, they would reap the whirlwind of peasant hatred. If the upper class acted out of good sense, they created a great deal of good will toward their class. The peasants were freed from the arbitrary terror of a bad manager and their relief at this occurrence would create larger surpluses for years. If any single act could have caused the serfs to love their masters, it would have been this one, even if it was performed not out of any abstract sense of justice, but out of simple self-interest. The feudal histories of England and Russia can be seen as examples of the extremes of good and bad management.

In England, for example, after the difficult Norman Conquest, the Normans displaced the Anglo-Saxon rulers, torturing and killing them. One expert estimates that about 20 percent of the population was killed, replaced by about half that number in conquering Normans and their allies, and the productivity of the land was cut in half for a generation.[8]

The Norman barons soon found that having a king was not all that much fun. After William the Conqueror, succeeding kings expanded their powers not only over the Angles and Saxons but over the Norman lords as well. Richard the First, for example, greatly increased taxes on the barons, reducing many of them to an economic condition not unlike that of their serfs. The serfs lived on the margin of starvation anyway, as the staple item in their diet was acorns; everything else went to their masters. But when John became king and threatened to do even worse to his fellow Normans, these Norman barons raised up an army against him. In order to obtain a large enough army to be effective, they had to get the peasants on their side and therefore they had to promise concessions to the Anglo-Saxon serfs, concessions that wound up in the Magna Carta, or Great Charter.[9] Curiously, the barons kept their word and instituted certain basic rights (they were listed as promises by the king) in the charter that pertained to all people—for example, the right to trial by jury. Additionally, since barons as a noble class were not permitted to engage in the disgusting practice of commerce, the conquered peoples did, and the Magna Carta provided several notable examples of the protections of Anglo-Saxon merchants in the sections in which the king made various promises for them. One concession promised that "the city of London shall have all its ancient liberties and free customs both by land and by water. Also we decree and agree that all the other cities, boroughs, villages and ports shall have all their liberties and free customs."[10] In short, freedom of trade for merchants.

In Russia, the opposite attitude on the part of the barons persisted down to the time of the revolution early in the twentieth century. The Russian army and the czars for centuries had to spend their time putting down the revolutions of bitterly disaffected peasants. Hence, when Germany invaded in the First World War, they found the Russian army full of peasant conscripts who were neither well trained nor interested in fighting. They were easy pickings for the well-trained Germans, and slightly later the democrats and then the Marxists found revolution was at last easy and wonderfully bloody. Many a noble line was utterly extinguished in that series of wars and revolutions. As long as the feudal system prevailed in Russia, it exhibited all the traits of incompetency that led to its demise. It was not functioning in a fashion that would preserve its people—it was not being used in accord with the requirements of survival.

The Duma, or Russian parliament, is one example of failure to deal with the changes required to preserve the system. It was not established until the reign of Alexander II, who ruled from 1855 to 1881. This is hundreds of years after England had a functioning parliament. Alexander, who set up the parliament, appears to have been a lonely fellow, almost by himself in understanding that it was necessary that the barons be restrained, and that for example they should not be allowed to simply kill peasants and then pay a small sum to the baron whose peasant died at their hands. It was Alexander who had begun programs to end the worst conditions: for instance, he tried to end household slavery,[11] although it is dubious his edicts had much effect, for in Russia, as a typical decentralized feudal state, the orders sent down by the central authority seldom had legal effect much beyond the outskirts of Moscow. The Russian anarchist, Prince Peter Kropotkin, bears witness to the barbarities practiced by the nobility, and he does not spare his own father in his revulsion at the treatment of the serfs. He reports that his father, stung by the gibe of another country noble that his serfs weren't producing children fast enough, simply ordered the marriages of all those of his serfs who were of remotely appropriate age, thereby ensuring a steady supply of serfs.[12]

Yet it must not be forgotten that through all this amalgamation, even when it was the vigorous and efficient amalgamation of England, that it was amalgamation to nationhood that was being accomplished. Equality of all persons was never on the agenda. The class structure remained rigidly in place. To take a minor example, as late as the time of Lewis Carroll (the second half of the nineteenth century), he could spoof the lower classes in his character of the messenger who reflected "Anglo-Saxon attitudes" in strange postures and gestures.[13]

In some nations, the structure became so rigid that further change from within became impossible. Everyone, rulers and ruled alike, would become so tied to traditional forms that only violence from outside could provide alteration, as the cases of Japan (think of the samurai) or India, at the times of their conquest, could testify. In Japan, of course, the samurai helped their own eventual conquerors by giving up the weapons—firearms—that might have allowed them able to defend themselves.[14] They drove firearms out of their society because any peasant, without much skill, could use one and bring down the most clever samurai, and that was simply unthinkable. But in their own foolishness they exemplified their station in society: the noble lords, the warriors, knew that the virtues of the military hero were the only true virtues, and that peasants, digging in the dirt and forbidden the possession of weapons, could not possess these virtues. The peasants were necessarily inferior.

Looking back, it is curious that while production was (and is) the fundamental requirement of life, unlike war, production and trade then had no virtue attached to it in the eyes of the rulers and perhaps in the view of all. Labor was base. As Thomas Aquinas would say, and this in the later stages of feudal development in Europe, "So trading, considered in and of itself, always implies a certain baseness, because it hasn't any honest or necessary object."[15]

Of course, a baron or lord might well have had some regard for his serfs, his inferiors; he may have even liked a few of them personally. Certainly he might have taken pride in his villagers, much as he would have appreciated a good horse, which is to say he would have appreciated them only in the lower station to which God had assigned them. In any case, he would ride both horse and serf in their own ways, making use of them for his purposes, not theirs. As Longfellow has one of his characters say in his *Hyperion*, "In this world a man must either be anvil or hammer."[16] This summarizes as well as can be done the relations of king to lord to peasant throughout the feudal period. It has its applications even today.

This conduct, in a word, is exploitation. Exploitation is simply the taking of the values (including the lives) of others without their consent, and using their values (or lives) for one's own purposes. It is the hallmark of the state, from earliest times to the present. It is what the state exists to do, that is, to allow some people the power and opportunity to exploit others. If we would put this in evolutionary terms, we would probably say that those who exploit cheat their societies and their fellow beings. In David Sloan Wilson's words, they "share the benefits without sharing the cost."[17] Those who exploit others do so because they believe it provides an evolutionary advantage to them. They will only cease to do so when they are destroyed, conquered and made to serve other exploiters, which are the human solutions so far. They will also cease to exploit when it clearly no longer provides an advantage to them, when their fellows simply will not allow the slackers and cheaters to live without labor and creation.[18]

Let me return to the development of the feudal state. It was rigid, we know, in its internal relationships. But in its clashes with the other feudal states around it, it showed a certain dynamism, as kings, barons, and other lords tried to expand their domains. While relations within the state remained static, while serfs remained in their status for generations, the outward relations of the state were such that they must grow or die. Oppenheimer is so worthy of quoting here that I cannot resist a lengthy passage:

> States are maintained in accordance with the same principles that called them into being. The primitive state is the creation of warlike robbery; and only by warlike robbery can it be preserved. The economic want of the master group has no

> limits; no man is sufficiently rich to satisfy his desires. The political means are turned on new groups of peasants not yet subjected, or new coasts yet unpilfered are sought out. The primitive state expands, until a collision takes place on the edge of the "sphere of interests" of another primitive state, which itself originated in precisely the same way. Then we have for the first time, in place of the warlike robbery heretofore carried on, true war in its narrower sense, since henceforth equally organized and disciplined masses are hurled at one another.[19]

A newly developing state, after all, does not exist in a vacuum, and kings and barons alike understood that they could gain wealth only through possessing more land and peasants. The amount one gained by careful maintenance, while useful, was small compared with what one gathered through conquests. And as the baron in state A understood this and wanted the land of a baron in state B, that feeling was totally reciprocated. To exploit others, it was felt, was good; others were inferior, were created by the god or gods to be exploited by the master class. This was as truly in the heart of the noble when he gazed covetously across the river at someone else's village as when he extracted his tribute from his own peasants. That other village, the noble felt, would shortly be his by the only right that existed, the right of might, of force, of violence. Trial by combat was god's law, and god would reward the virtuous man with victory. This baron was that man, and he was impelled to war. The skulls of the lords of that land would be his drinking goblets and their servants would be the servants of his servants.

As he turned to practical matters, he noted that his body of retainers was large and well trained—so large that it used most of the surplus provided by the peasants, costing him dearly—and that harvests had been down the past couple of years. His land was taking a turn for the worse, his income was off, but just over the river was a village that would recoup his fortunes. So the bugle sounded, his knights mounted up, the spears were readied, the warriors commended their souls to their god, and the invasion force moved to battle. At last the warriors experienced the great peacefulness that comes from fulfilling their function well, of achieving their destiny. And the soldiers who returned from battle found they had been changed by it: killing others in that fashion was the most meaningful thing they could have done. It was an "experience transcendent to all other meanings."[20]

It would be pointless to criticize this. Today it is generally felt that aggression is bad, that is, aggression in the sense of the violent use of others for one's own ends. Indeed, so widespread is this feeling that sometimes even the United Nations passes resolutions condemning aggression. Often this is mere hypocrisy, as nations just finished with their latest blood quest point with alarm at others just starting out. Even so, by their very hypocrisy, leaders of nations recognize that there is a sense in the air; many people now understand that there is something wrong with violent aggression. The horror with which we turn from terrorism—the use of "collateral damage" to achieve one's goals—and ethnic cleansing by tribes and nations in pursuit of ancient hatreds indicates this as well. Most people are different from the way they were a thousand years ago. As we change, little by little, we move to useful ways of conducting ourselves. Bloodshed and rapine are not generally seen as beneficial as they were then. We begin to recognize these are not useful survival mechanisms.

Among feudal peoples no such thoughts of outrage are likely. To blame the class of warriors, kings and vassals alike, for being violently aggressive had no more sense to it then than blaming a rose for having thorns. It was felt to be a natural thing. Aggression in the masters was a good quality, to invade another's territory was simply to have the good sense to strike before one was struck.

These clashes, these small wars, caused the growth of the state "internationally." State A conquered B, C, and D, formed an alliance by marriage with E, conquered F and was in turn conquered by G, which then went into battle with H that had undergone a similar process of expansion itself. In this expansion, every form of irregularity was exhibited and every device imaginable was developed to expand and preserve the power of the noble class and the central ruler.

The regularity that is found is that every successful state would develop to a point beyond which it could not expand further without losing part of what it had already acquired. In effect, centralizing the state led to a point when that process was reversed and the state was decentralized. The central monarch controlled his court and the immediately surrounding lands thoroughly, but the farther reaches of his lands had to be run by his lords and vassals, and the farther away they were the more actual power was maintained in their hands. Even if the monarch was blessed with level land, good roads, and an army equipped for speedy travel, this would be true, but even in that case decentralization at the edges of empire would be the rule. Even today in feudal states such as Somalia or Afghanistan, the local nobility—we call them "warlords" today—run the country beyond the reach of the official leader of the state. Sometimes, as in feudal France, this would be less evident. In other cases, as in Russia or China, decentralization would be carried to extremes, and empires would form, decay from the outside in, and then die out and be reborn under other noble classes and monarchs. Decentralization was always a threat to the overall rule and power of the king, emperor, or shah.

Why should this continual movement to decentralization be so commonplace? Remember that war was a common occurrence between feudal states. In order to survive, a monarch would have to have strong and effective leaders in his border areas.

Imagine yourself in the position of a feudal king. Consider things from the expertise of Machiavelli, the greatest strategist of feudal warfare, who advised that a king must think only of war if he is to survive.[21] You looked out upon what you knew as your country. You saw that your farthest border was a good ten days ride on a fast horse. That border was a river and across the river was a rival kingdom. If you centralized your armed forces and then if the king across the river attacked, your baron in charge there could only send a rider to notify you, and you would have lost. Why? Ten days to learn of the attack, another few weeks to round up your troops and provision them, and then rather more than ten days, perhaps twenty (you would have foot soldiers too), to get to the scene of the attack. Your opponent would have had two or three months to entrench himself and get ready for your counterattack. If he had waited to attack until very late fall, you probably could not have responded at all that year—moving troops through bitter cold and snow was seldom possible—and then in the spring, your troops would not become available until after the spring planting is done as your nobles would have to supervise this. What hope would you have had of reclaiming your territory?

In short, this strategy led to defeat. You needed to have strong and effective leaders in your border areas, and they would have to have had access to the means of equipping fighting forces. Therefore you, the monarch, would have had to turn over much of your power to the locals. So your border lords would thus increase their military strength at your expense, as the cost would be borne, in effect, by less surplus being sent your way. These border leaders would become more and more independent. They would still swear allegiance and fealty to you, but in practice they would have become substantially independent of your central authority. In order to protect your domains you had armed those who would eventually desert you and assert their own authority. Out of necessity, you would have engineered your own demise. Oppenheimer here comments regarding these barons on the outskirts of things:

> The crown must pay more and more for their services, and must gradually confirm them in all the sovereign powers of the state, or else permit their usurpation of these powers after they have seized them one after the other. Such are the heritability of fiefs, tolls on highways and commerce (in a later stage the right of coinage), high and low justice, the right to exact for private gain the public duty of repair of ways and bridges . . . and the disposal of the military services of the freemen of the country.[22]

So the feudal lords in this position eventually would come to the conclusion that they should be independent of their king. After all, why should they send a part of their peasants' surplus to a faraway king who did not protect them, but in fact *they* protected *him*. What benefits did they gain? When the answer was clearly none, and when the chance presented itself, they revolted and established their own little kingdoms. They had the power to do so.

Crucial to the entire development of a feudal state was the master's possession of the land and control of the peasants who farmed the land. Kings might have asserted ownership of the land and the serfs, the labor motors who worked the land, but the ones who actually held the day-to-day power were the barons. And key to increasing wealth was successful warfare where the winning lords took more land and peasants. Since wealth was regarded as generally static, the only way one got more was to take the land and peasants away from others.

Hence the importance of the political means, that is, the power to exploit the serfs or peasants. This was why the nobility were not interested in economics. And that is why, when countries like Spain explored the "new world" and colonized it, they gave large estates to their nobles, who enslaved natives, who became the serfs for their conquerors. Then, when these people in many places nearly or completely died out from overwork and disease, the Spaniards imported slaves from Africa to work for them. If you went to the Americas for your king, you expected to take over land and peasants to become wealthy. Who goes somewhere new to take a vow of poverty? Only a few monks who often complained about the conditions of the slaves.

Who cared about economics? Who cared about production? Not the nobility, who only knew they needed land and peasants. Those who did care about economics, one of the minor interests in life, were the serfs themselves and the occasional freeman who served as foreman on some estate: the people who must somehow cobble things together and hang on.

The upshot of the feudal way of life is summed up by Oppenheimer as follows: "Thus there starts from the fields, whose peasantry support and nourish all, and mounts up to the 'king of heaven' an artificially graded order of ranks, which constricts so absolutely all the life of the state, that according to custom and law neither a bit of land nor a man can be understood unless within its fold."[23]

It was an orderly society that, if it were not for foreign affairs, that is, for war, would become and remain utterly stagnant for centuries, moving from centralization through rebellion to decentralization and then through war back to centralization, without any forward movement at all. The peasants at the bottom had no stake in who was baron or whether he was baron or king. They and their descendants, as far as they could see, were bound to the baron and his descendants forever. Their status did not change, although one or another baron or king might rule them. In some feudal societies, the classes and gradations of classes were so thoroughly developed that they became fixed castes, which then slowly degenerated until they were blown apart by some great outside force, as happened when the British conquered India. That invasion and conquest caused the long-term, indeed, fatal damage to the castes, a process of dissolution so slow that it continues even today in otherwise modern India. But it is not surprising that those who like law and order, calmness and obedience, the certitude of decay, find their ideals in feudal times.

As a final note, if may be suggested that while I have talked much about the barons in their many ways and guises, I have talked little about the monarchs. This is because while monarchs technically sat at the top of the pyramid, they frequently had less power than the barons they allegedly ruled. If I need to make amends to monarchs, let me do so by way of the story of Oleh, son of Vselslav, ruler of Kiev for part of the tenth century. I give you a small part of an epic poem about him, which reports that when he was five years old, he went

> over the earth.
> Mother Earth trembled;
> The wild beasts in the forests fled;
> The birds under the clouds flew away.[24]

If kings did not actually control as much as they thought they did, they could at least bask in the glories of poetry and the declamations of the court bard. Feudalism had its pleasing illusions.

NOTES

1. G. Molinari, *The Society of To-Morrow,* translated by P. H. Lee Warner, introduced by Hodgson Pratt, appendix by Edward Atkinson (New York: G. P. Putnam's Sons, 1904), pp. 1–2.

2. Franz Oppenheimer, *The State,* translated by John Gitterman, introduced by C. Hamilton, http://www.franz-oppenheimer.de (last accessed June 14, 2007). See especially chapter 5.

3. Charles Howard McIlwain, *The Growth of Political Thought in the West* (New York: Macmillan Company, 1932), p. 182.

4. Jared Diamond, *Guns, Germs, and Steel* (New York: Norton, 2005), p. 276 makes the point that the elite in this kind of society logically uses its arms to provide "public order" and limit violence. Note this applies to violence even by its own members, which we will get to in a bit.

5. Richard Wrangham and Dale Peterson, *Demonic Males: Apes and the Origins of Human Violence* (London: Bloomsbury, 1997), p. 77.

6. Diamond, *Guns, Germs, and Steel.*

7. David Riesman, Nathan Glazer, and Reuel Denney, *The Lonely Crowd* (Garden City, NY: Doubleday, 1953). In particular note their discussion of the "tradition-oriented" individual and their comment on page 33.

8. David Howarth, *1066: The Year of the Conquest* (New York: Barnes & Noble, 1993), especially pp. 197–200.

9. Certainly the barons would have preferred to get rid of John altogether. He had a modest empire on the continent, and his expenditures to maintain that empire were extraordinary. What baron would care about this empire?

10. See Samuel E. Thorne et al., *The Great Charter* (New York: New American Library, 1966), chap. 13. The translation is by the author. Note that McIlwain argues that this kind of treatment of different classes is inherent in feudalism. See Charles Howard McIlwain, *The Growth of Political Thought in the West* (New York: Macmillan Company, 1932), particularly chapter 5, "The Early Middle Ages."

11. G. D. B. Gray, *Soviet Land* (London: Adams and Charles Black, 1947), p. 131.

12. See his *Memoirs of a Revolutionist* (New York: Houghton-Mifflin, 1899), pp. 52–54.

13. Lewis Carroll, *Alice in Wonderland and Through the Looking Glass* (Chicago: John C. Winston, 1923), chap. 7.

14. See the story as told by John Hillman, *A Terrible Love of War* (New York: Penguin Press, 2004), pp. 161–64. He points out the position of equity that firearms bring about between a "lowly peasant" and a "noble lord," p. 162.

15. Thomas Aquinas, *Selected Political Writings,* trans. J. G. Dawson, ed. A. P. D'Entreves (Oxford: Basil Blackwell, 1965), from the *Summa Theologica,* question 77, article 4.

16. Henry Wadsworth Longfellow, *Hyperion: A Romance* (Boston: Houghton, Mifflin and Company, 1893), p. 371.

17. David Sloan Wilson, *Evolution for Everyone* (New York: Delacorte Press, 2007), p. 129.

18. One well-told story suggesting one way this might happen in our future is the fable of "Idle Jack" in Eric Frank Russell's wonderful old science fiction novel *The Great Explosion* (New York: Avon, 1962), pp. 140–42. It is an excellent tale of how refusing to be exploited can work, of how, in evolutionary terms, the cheater can be made to do his or her fair share without any use of physical force being required.

19. Oppenheimer, *The State,* p. 41.

20. Hillman, *Terrible Love of War,* p. 10.

21. Machiavelli, *The Prince,* trans. Luigi Ricci, rev. E. R. P. Vincent (New York: Mentor Books, 1952), chap. 14.

22. Oppenheimer, *The State,* p. 74.

23. Ibid., p. 83.

24. As lovingly quoted—there is a great deal more of it, of course—in Michael Hrushevsky, *A History of Ukraine* (Hamden, CT: Archon Books, 1970), p. 48. The work is an excellent and sympathetic study of that troubled and newly reemergent land.

CHAPTER 3

MARITIME PEOPLE

Britain, in many ways the greatest Maritime state, once ruled the waves: since then it has simply waived the rules and gone any way it can.

The maritime state, with its very different path of development, clearly counts as an early form of state, establishing itself at approximately the same time as feudal states make their appearances, although this varies greatly because many maritime states were only operational that way for part of their existence.[1] This kind of state was founded by the seaside, or occasionally at some point where rivers flow into each other. One would quickly think of the Vikings as having these kinds of states. England, of course, became one and changed history greatly because of it. The island peoples of Asia not uncommonly established this sort of state as well.

The distinctive feature that created this state is that these primitive peoples frequently found it beneficial, from the earliest stages of their development, to venture forth in ships, stealing from some places while trading in others that were well enough defended that stealing was impractical. Henri Pirenne, for instance, speaks of piracy occurring in Europe, with that practice being the "initiator of maritime trade among the Greeks . . . as among the Norse vikings."[2] The same impulses that motivated kings and barons were at work here, only the conditions were different. It is unclear just how important evolutionary change has been in these situations, but a recent study on mammals in general indicates a "dramatic" change in shape and sizes in island mammals.[3] Other kinds of quickly accomplished changes may well exist, so the different conditions of living, and hence, improved survival strategies, may help trigger the appropriate changes in genetics, as well as being results of these changes. The maritime peoples adjusted and the results were very different from what might otherwise have happened.

Flowing from those changed circumstances, the maritime state was usually much more centralized than the feudal landed state, for the obvious reason that it was not spread out over a great area, but concentrated upon its major trading harbor, such as would be found in one of the fjords of a Viking state. Indeed, so closely associated are the Vikings with this kind of state that a few comments on a typical Viking state may be useful.

These people originated along the narrow inlets in what are now Norway, Sweden, and Denmark. The areas they inhabited were limited by the mountains in the interior. Since most of these people were quite isolated over land, they took to the sea. The leaders of the society, the jarls, or earls—the upper class—did not farm, and there was not much game to hunt, so before 800 CE they were sailing down the fjords and going off to raid and trade. Their slaves, or thralls, were left behind to do the limited amount of farming possible, supervised by a small class of freeman and by the noblewomen. The first recorded Viking strike was against a monastery on the eastern coast of England in 793 CE. Their attacks on England occurred regularly for two hundred years. So familiar did they become with the English that they even put down a few settlements there. They also hit Ireland and conquered most of it, attacked in France and Spain, and sailed into the Mediterranean and Caspian seas. Some tribes moved down the rivers into Russia; one state even attacked Constantinople. They sailed west as well and established colonies in Iceland, Greenland, and briefly, the North American continent. When they sailed on these colonizing expeditions, they took women and slaves and settled down to rule their new lands.[4]

Much of this would remind a student of early Greek history how little states like Athens did exactly the same thing. Viking colonies, like those of the Greek states, were put down because they so easily overpopulated their small homelands, not having the agricultural land to expand over. Later on, when England became a maritime state, we find the same sort of pattern, but of course a worldwide empire was created. I suppose one of the earliest points of English experience with developing the maritime life was early in the sixteenth century, when they began fleet building, the early high point being the voyages of the pirate Francis Drake, who divided his time between stealing the gold from the Spaniards that they had stolen first from the Indians in their empire, and exploring around the globe. It was the English merchants who pushed these developments, and when, as in North America, they planted colonies, it was with the understanding that as commercial ventures, the colonials would provide England with raw materials but the manufacturing would remain in England. There was also the defeat of the Spanish Armada in

1588, which was a key event in setting England on its maritime course. By 1600, the East India Company, the most famous of the merchant companies, was established. The common term for this kind of domination was *mercantilism;*[5] although today we might call it an early variety of *crony capitalism,* for those with "pull" with the monarch got the plums, or such as they could acquire in dealing with the merchants. One need only look at the colonial charters for North America to see how this worked. Read, for example, William Bradford's report (beginning in 1608) about the establishment of the Plymouth colony to see how the leaders of the group that wanted to come to America had to deal with treacherous agents of the king, some seriously incompetent crony capitalists, and how they finally managed to cobble together a deal.[6]

Wealth in a maritime state in its original manifestation was gold and precious jewels. The jarls wanted cash, so to speak.[7] On occasion when they colonized, they turned into earls and barons, with land and slaves occupying their attention. But the true maritimer wanted movable wealth since he was indeed always moving about. True, he would own a bit of land back home and keep slaves to work it, but he paid it little mind as long as he could eat when he returned home after his satisfying excursion into the broader world of piracy and trade. This emphasis upon movable wealth was developed by the English into an industrial powerhouse for the homeland, far outstripping the simple form of movable wealth that the Spanish had enjoyed and yet misused because they acquired it through land and slaves, and thus did not know how to properly manage it.

Indeed the English were far more efficient in exploiting their colonies than the Spanish, perhaps because the Spanish merely transferred a quite brutal feudal system overseas and never really understood what they were doing. Despite, or because of, the systems of forced labor that the Spanish instituted and the incredible theft of gold and silver from the

Indians that the Spanish practiced,[8] the wastrel Spanish lords were never able to defend their empire well in the face of serious danger, as Napoleon demonstrated to the world at the beginning of the nineteenth century by conquering the Spanish and putting his brother on the throne.

But what about the English? By the early nineteenth century, London was the world financial capital, where credit was extended, often on a political basis, and commodities were speculated on. London financiers and indirectly, the British government, controlled much of the world's monetary system through the operations of the Bank of England. To give you a small sampling of how far their direct colonial system extended, let me offer you a list. The list includes perhaps half the conquests and takeovers and none of the places just strongly influenced. The dates of conquest are approximate. The chronological list ranges from Antigua to Zimbabwe, from a conquest in 1627 to one in 1920, from islands to continents, from Africa to the Caribbean to the Middle East and the Nile Valley. The old boast about the sun never setting on the British Empire was perfectly true:

Barbados	1627
Antigua	1632
Jamaica	1655
Canada	1763
Australia	1770
Grenada	1784
Dominica	1805
Mauritania	1810
Malta	1814
Guyana	1815
Singapore	1819
India	1828
New Zealand	1840
Bahrain	1861
Nigeria	1861
Malaysia	1867
Lesotho	1868
Egypt	1882
Botswana	1886
Brunei	1888
Malawi	1891
Uganda	1894
Zimbabwe	1897
Kuwait	1899
South Africa	1902
Swaziland	1903
Jordan	1920

The infinitely shrunken empire still controls Wales, Scotland, Northern Ireland, the Falkland Islands, Pitcairn Island, and so on. Much of the story that remains is sordid by contemporary standards: for example, in 2006 it became known that the peoples of the Chagos Archipelago, forced out of their land in 1967 by the British government (an act hidden by royal prerogative), were displaced in order for the British to make a deal with the United States. Britain got a price cut on the Polaris missile system and the United States got a base on Diego Garcia. These people perhaps will finally go home, as the High Court in London has recently ruled in their favor.[9]

This brief discussion of the British mercantilism cannot even begin to touch upon the raw materials that went to England to enrich the people there. An entire book could easily be devoted to the subject.

All these things meant that a maritime society was extremely different from the feudal state. Under feudalism, change was rare and hard to come by. But with the maritime pirates/merchants, change could take place rapidly. Sail out with a good captain and have a few profitable trips, and your share of the booty could let you buy a ship of your own, an obvious improvement in status. Get to the point where you owned four or five ships, and you were a magnate. But then came another season and you fared badly. Your ship limped into port after losing an almost fatal encounter with a ship from another people. Maritimers killed each other just as enthusiastically as they killed people in landed states, for there was no professional courtesy among pirates.

You vainly waited until the harbor closed for the winter and your other ships did not return at all. Your status took a nosedive, and while you would never fall far enough to become a thrall or common freeman, you would have suffered a grievous loss. In short, with the maritime state, there was a good deal of social mobility. England exemplified it, as one thinks of the merchants, the descendants of serfs, who traded around the world, set the currency standards for the eighteenth and nineteenth centuries, and became, many of them, incredibly wealthy through the means and advantages this kind of state affords.

There was even a possibility of social mobility for a slave in a system such as the Vikings had: one could move up to freeman status. If a slave had a decent master, he would let the slave farm a tiny section of his land for himself and keep what was earned off it. In time, the slave might save enough to buy freedom. It was also possible for a master to simply set someone free for long and devoted service. The feudal state, which depends so much more heavily on slavery, could not allow this opportunity except on extraordinary occasions, if at all.

But still, as a slave, one had no more status than a cow or a pig. The slave's hair was cut short, and he or she wore a heavy collar around the neck. If someone else's master killed the slave, he had to pay a fee to the owner, because the slave was property, no more.

Since a maritime state is necessarily more exciting than a landed state, with open ports and sailors going to very different areas and always learning new things, it must have been correspondingly harder to maintain slavery. Slavery among the Vikings had to be bolstered by their religious propaganda, which no doubt eased many a dubious conscience of a jarl or freeman who found new ways of doing things and occasionally, new friends abroad, or slaves, or just people to kill. Heimdall, it is said, created people, and he first created the slaves, who are described as dark with dull eyes and gnarly fingers—ugly, and yet strong enough to bear all the burdens imposed on them. The first thrall couple created quite a litter. The children were given such unpleasing names as, for the boys, "Howler, Stumpy, Bastard, Sluggard, Lout," with the females named "Drudge, Slattern and Dumpy,"[10] among others even more obnoxious. It is an axiom of propaganda that you use demeaning names to describe those whom it is in your interest to demean, and this facilitates your power over them. One thinks perhaps of the Nazi use of such items as the *Protocols of Zion* to help them dehumanize the Jews. And one needs to remember that this ties back into tribalism and the view of one's own tribe as the only really human group, or as today, the terms soldiers give to the enemy: *gooks*, for example, as the US troops called the enemy in Vietnam, or *haji* in the war in Iraq. The others, the strangers, were and still are evil impostors, pretending to be human.[11]

It may be noted here that the maritime state does not have amalgamation as one of its great emphases. While there are exceptions, it still remains general practice that when, for instance, an ancient Greek state like Athens colonized, it sent its own people out to settle and enslave, displace, or otherwise eliminate the people already there. When the English came to America, they did not amalgamate with the Indians—they drove them off and killed them. Their conduct looked very typically maritime with feudal overtones. By 1645, the Plymouth Colony had begun requiring the Indians to pay tribute to them. The colonizers even took Indian children as hostages to enforce their orders upon these inferior beings.[12]

By contrast, when Spain made a colony in South America, they ruled it by the power and authority of the "Spaniards from Spain," and any illegitimate children were regarded as peasants, or as *mestizos*, of mixed and thus inferior parentage. The Spanish followed their own precedents, as well as those of the Inca or Aztec empires, which, when they conquered surrounding tribes, took these people as slaves and used many of them as sacrifices to the gods. The Spanish, of course, passed up the sacrifices and just worked the conquered Indians to death.[13]

When the maritime Vikings colonized in York, they did not amalgamate with the Angles and Saxons for centuries. Many times (and this is quite typically maritime) there was no attempt at colonization. When England conquered India, they sent administrators and the army to manage such items as needed to be handled for England's benefit, and when they conquered Ireland they ran that nation as absentee landlords would, with English gentry receiving rental payment from Irish peasants, even though the gentry might not visit "their" possessions for generations.

A maritime state is always subject to change: follow movable capital (or industrial opportunity) and it can take you to places you never imagined, with alterations in your own life you could not have conceived beforehand.

The pepper trade is a famous example. Pepper was one of the first products dealt with in international trade, having been a subject of "piracy and trade" as far back as four thousand years ago[14] in the trade between India and Southeast Asia. For hundreds of years, it was as valuable as gold and was traded like gold in many places. At one point the trade was an Arab monopoly, although due to the profits others were not content to leave it in those hands. By the second or third century, the Greeks and then the Romans made voyages to India after pepper. In time they were displaced by the Spanish and then the Portuguese. Indeed, at one time Lisbon was the center of the trade and was the wealthiest city in the world because of it. The Dutch then took over and finally the British. Even the Italians from such merchant cities as Venice and Genoa were involved for a time. Today, all manner of companies and nationalities engage in the trade, which still seems to involve about one third of the value of all spices. I was recently looking at a spice catalog,[15] and noted that today we have available Sarawak from Malaysia, Tellicherry and Malabar from Kerala and the southwest coast of India, Muntok from Indonesia, and Szechuan from western China. The best pepper mills appear to come from Germany. Those who today go to these places and make their economic deals represent the best of the maritime traditions of learning and giving learning, a largely unintended consequence, to the peoples they contact.

I don't know if the pepper trade is a great changer of attitudes today, but certainly it has been so in the past. Imagine in the fourteenth century sailing all the way from Europe to India in search of pepper! What amazing things one must have seen on the way—things very unsettling to any person of stolid habits and established ways of being. People with very different lifestyles, religion, dress, manners—people different in every way, who had a product one craved, and who had to be dealt with. What wonders to be seen, what glories to plunder! It was therefore eye-opening and mind-shaking for all the parties involved. Pepper spiced people's lives in more ways than the obvious.

"The game's afoot, Watson!" we might cry if this were a detective novel, and perhaps this can be considered a kind of detective story, because with all this talk about change, you may reasonably deduce that the very form of state would shortly be altering to accommodate change.

And if one looked at the matter from the standpoint of the evolution of human societies, it is clear—in a very general way—what happens. We started with very simple tribal societies and moved to feudal states with a complexity of classes and customs assigned to each class, with some legal, if seldom-enforceable, rights attached to them. We saw other states "go maritime," branch off and explore the mysterious ocean or travel down the river, there to find far more complex puzzles than people had ever had to solve before. It is evident by now that the simple response "my tribe is good, all others are bad," or "I am friends with my people in my tribe, but all others deserve to be killed or enslaved, something that I will do with great enjoyment as soon as I can," will no longer work efficiently or well. Tribalism has long since outlived its usefulness. Its uselessness was a great deal more apparent to people in maritime states, as in the course of their travels they spread difference and variety quite as much as they were affected by it. These changes would have penetrated the skulls of all but the most obtuse baron or king. A new kind of person was called for, one who could handle diversity. A premium was put on new abilities and adaptability to change, and usually this called forth the appropriate talent. Out of the clash of foreign affairs—that is, war and trade—for both maritime and landed states, came a great development: the city and the people who inhabit it.

But before we talk about city people, let me offer one last comment, not so much on the maritime state itself, but on the opportunities people have to choose and the consequences of choice in these circumstances. We cannot choose everything, but there are many choices we can make. Let me tell a story about China and Zheng He.

At one time the Chinese had started working on having a maritime state. They had compasses and all sorts of innovations to make it possible. When China had disposed of the threat of the Mongols, the Ming dynasty decided to take to the sea. In the late fourteenth century, the emperor wanted a navy. He ordered a shipyard built, had forests of appropriate trees planted for lumber to make the ships, and even established a school to train interpreters. Zheng He was appointed to be the admiral for this fleet, and he was to collect tribute from all the tribes and peoples who were not fortunate enough to be Chinese. The first of seven voyages got under way in 1405. These voyages would take the admiral and his sailors over thirty-five thousand miles. He had a massive fleet for the first voyage, sixty-two huge ships over six hundred feet long (some estimates say only four hundred feet). There may have been many other small ships as well. By way of contrast, Columbus left on his famous voyage with only three boats, the longest of which, the *Santa Maria*, was seventy-five feet. On the way out, the Chinese fleet—the Treasure Fleet—went to Vietnam and India, and coming back, to Sumatra and Java. Six other voyages were undertaken and the admiral began an aggressive trading program with warehouses and military outposts established in many different countries. Ambassadors from many places were brought back to China. Foreign kings, if they did not display the appropriate respect, were disciplined. Voyages went to the Middle East (the admiral, a Muslim, got to visit Mecca) and much of the African coast was explored. In short, the Chinese were rapidly building up a major commercial and governmental venture, and the vast wealth and power that was beginning to accrue heralded the beginning of a major maritime state. But far more than in the English state, all depended on the ruler and his court. In China, the court wound up rejecting the maritime state, understanding that it would lead to great change and liberalization, which violated all the ideals of the landed feudal state and all their familiar rules and powers of exploitation. After the admiral and the emperor died, there was a new emperor, but no new admiral. The new advisors, the court, of the new emperor argued before him and won the battle against the maritime effort. The ships were destroyed, the factories where they were made were destroyed, and edicts were issued banning the construction of new ships above the level of coastal fishing boats. Even Zheng He's logs were destroyed. Every effort was made to erase the wealth and the changes that his efforts had begun to make.[16]

Imagine if those maritime efforts had continued. Perhaps they were more difficult than they would have been for an island (quick evolutionary change would not be such as an assistance here), but one would hardly have picked the Spanish as a maritime people either, which of course they were only sporadically and badly. The later Portuguese, Spanish, and English expeditions would have run into an extraordinarily powerful Chinese empire. Possibly the Chinese would have expanded into the North and South American continents (Zheng He, by one account, touched Australia). China would have had an empire of incredible size. Would there even have been a British Empire? Would Americans, instead of speaking English and Spanish, be speaking Mandarin today? The circle of choices we have may sometimes seem narrow, but a few royal advisors, centuries ago, made a choice, based on their own interests and tribal/feudal concerns, that has had worldwide consequences. It may be that someday we will have a choice beyond the myriad of choices in our daily lives: Will we do as poorly as the court of the emperor?

NOTES

1. See Franz Oppenheimer, *The State,* trans. John Gitterman, introduced by C. Hamilton, http://www.franz-oppenheimer.de (accessed June 14, 2007), chap. 4.

2. Henri Pirenne, *Medieval Cities,* trans. Frank D. Halsey (Garden City, NY: Doubleday, 1925), pp. 75–76.

3. Virginie Millen, "Morphological Evolution Is Accelerated among Island Mammals," *PLoS Biology* 4, no. 10 (2006): e321.

4. For example, see KRG Pendlesonn, *The Vikings* (New York: Mayflower Books, 1980), especially the chapters on "The Viking Onslaught" and "Viking Colonization and Exploration."

5. The comments by Ludwig von Mises, *Human Action,* 3rd rev. ed. (Chicago: Henry Regnery, 1966), pp. 664–66 may be a helpful start in understanding the mercantilist system.

6. William Bradford, *The History of Plymouth Colony,* introduction by George Willison (Roslyn, NY: Walter J. Black, 1948). See in particular the agreement that the colonists started with, pp. 49–51. Of course there were dissenters; some of the nobility of England, already rich in the feudal way, were more interested in establishing the system of landed estates in the "new world" than in doing anything that would benefit merchants. See the editor's note p. 48.

7. Merchants do best in maritime states, having more in common with the kings and barons there than elsewhere. And merchants do like their cash. For example, in the Magna Carta of England one sees the tide of commercialism starting to advance in the country even at that early date. The king's officials cannot take "grain or other supplies from anyone, without immediately paying for them," unless they can get the seller to agree to something else. In effect, in God we trust, but the king pays cash! Samuel E. Thorne et al., *The Great Charter* (New York: New American Library, 1966), my translation, chap. 28.

8. See Alonzo de Zorita, *Life and Labor in Ancient Mexico,* trans. Benjamin Keen (New Brunswick, NJ: Rutgers University Press, 1963), and Alfonso Teja Zabre, *Guide to the History of Mexico* (Mexico City: Press of the Ministry of Foreign Affairs, 1935). Perhaps one-fifth of all the gold in the world was taken by Spain from the mines of Mexico alone; and the figures of Zorita estimate that from a native population of almost seventeen million in 1532, the population in just seventy years dropped to just over a million people.

9. Said the High Court, the idea that a government official, simply on his own, could expel a whole people from their home, try to hide it, and all the while claim this was done for the "peace, order and good government" of these people, was "repugnant." Neil Tweedle, "Britain Shamed as Exiles of the Chagos Islands Win the Right to Go Home," *Daily Telegraph,* May 12, 2006, http://www.telegraph.co.uk (accessed May 13, 2006).

10. See the curious Viking verse at http://www.Vikinganswerlady.com (accessed May 27, 2006).

11. For a look at how the hatred of others shaped many of the mass movements of the twentieth century, see chapter 14, "Unifying Agents," in Eric Hoffer, *The True Believer* (New York: Mentor Books, 1958).

12. Bradford, *History of Plymouth Colony,* chap. 25, pp. 417–18.

13. Zorita, *Life and Labor in Ancient Mexico.* See also Jared Diamond, *Guns, Germs, and Steel* (New York: Norton, 2005), pp. 197–206 and elsewhere for a discussion of how the introduction of germs by the Spanish decimated the native populations in many places.

14. http://www.sallysplace.com (accessed February 4, 2006).

15. The Penzey's catalog is my source here, http://www.penzeys.com. I'll talk more about this process when we get to globalization, for obvious reasons. (Accessed June 2, 2007.)

16. There are a number of Web sites that mention this episode in history. See, for example, http://planet.time.net.my/CentralMarket/melaka101/chengho.htm. Also http://web3.asia1.com.sg/trip/journey/travel/china/ecread3.html and for a brief summary, http://library.thinkquest.org/20176/chengho.htm (accessed all sites October 1, 2006). His name in English can be spelled in several different ways. Note, by the way, that the admiral began his life in Yunan, and was captured and castrated when the Chinese army conquered his area. He was a slave, and as a favorite of the emperor's son, rose with his master. Castration was a way of making sure that your slaves don't spread new ideas, although it seems not to have worked in this case. It is a practice also designed to keep the gene pool as stagnant as possible, and as such is an early form of forced sterilization, among other methods used to keep alleged inferiors, in this case, non-Chinese, from breeding.

Part 2

THE PRESENT

CHAPTER 4

CITY PEOPLE

Societies and civilizations in which the cities stagnate don't develop and flourish further. They deteriorate.[1]
Jane Jacobs

We get at last to the city and the kind of people who populate it. Cities, in the sense of mere agglomerations of people, or in the case of Rome, a large collection of those who mainly profited from the rapacity and plundering of the empire, have been around for some time. The form and purpose of the city are now developing in new and very different ways. The city is becoming what it must be for progress to occur, something Oppenheimer calls the "industrial city."[2]

Speaking in very general terms, we can trace the growth of this new kind of city in these ways: In the beginning (that is, up to this point in our story), we had either maritime or landed states. In the first, the rulers accumulated wealth by piracy or trade, in the latter, by robbery. In both cases, exploitation was basic—tempered in the first, quite unabashedly triumphant in the second. The transformation from the usually quite small maritime state, already centralized, to a more complex structure that depends more and more upon trade has a distinct impact upon any more rigid feudal state it would come in contact with. It would make sense for the transformation to occur first on the maritime side, and this is usually the case. It works like this: First we have to understand with Oppenheimer and Jane Jacobs, the economist, that we are no longer talking so much about states and state development and state growth as we are talking about cities.[3] We are talking more about economic growth and economic development and therefore less about political growth and development. Henry Pirenne, for Europe, talks about how cities grow because of and out of trade, as that continent recovered, finally, from its long sleep under the empire of Charlemagne, a time when it was "a closed state. A state without foreign markets, living in a condition of almost complete isolation."[4]

Our whole perspective needs to shift. Exploitation is no longer the central focus, although it does remain in a less prominent position. We are beginning to talk about and emphasize freedom and all it brings in its train. The individual will no longer be so concerned about pleasing his master, but about pleasing his customers and pleasing himself or herself as well, a kind of genuine cooperation for mutual benefit. The complexity of these kinds of relationships will continue to develop over time.

So what happens? Take the case of the feudal state, decentralized as we know. Somewhere within the state, on some manor, more likely either one sufficiently far within the state to avoid spending all the time worrying about defense, or one where there has been contact with a maritime state and some of the maritime ideas have penetrated, the local baron comes up with a brilliant idea: the total amount of wealth is not fixed. Production can increase wealth. Economics is not a static study, but a dynamic one. The baron recognizes that it would be an obvious advantage to him to have that increase take place on his estate. So he decides to spur it on by announcing in some fashion a new deal to his peasants. "Henceforth," he might declare to them, "I will not be taking all your surplus. I will take (for example) one half of everything you produce. You may keep one half of all your earnings."

Peasants probably heard this with considerable doubt.[5] It was hard to imagine that one's baron wished one well and that a baron would actually keep his word to them. So for that first year, most of the peasants probably went on producing only as much as they had to; they were by no means a shiftless lot, but there was never any occasion to make more, when the "more" went just to the baron. But perhaps two or three peasants decided to take the gamble. They worked hard all that spring, sowing and planting, and all summer long, weeding and fertilizing. Come harvest time, they took their much larger crop to the mill for grinding. The baron or his overseer would be there. Pleased at the increase in production, the peasants who worked hard were indeed allowed to keep their full share of the greatly increased amount. Those peasants had enough food to live well during the winter and enough fuel to keep themselves and the livestock that shared their quarters warm. All the other peasants, half starving and half freezing in their miserable huts, vowed to do better next year. In a few places though, they probably got sufficiently jealous to kill the successful ones and steal the goods of the brand-new entrepreneurs. But come next spring, many more would be out working harder than they had ever done. More would be produced, the baron would see himself becoming wealthier than his fellows, and the peasants would be moving to a higher level of comfort. For the first time the peasants had an economic incentive to work, to become more productive, even to invent laborsaving devices, something no one had bothered with before. For instance, some peasants invented a new kind of harness for the draft animals. The old one, at least in Europe, had a strap across the throat of the horse or mule, which limited the pulling power of the animal. Drop the strap to the chest, the animal pulls harder, more work is done, more land is farmed, more crops are grown.

Most of the peasants now had more, and not all of it went into consumption. Some peasants some of the time used it as a capital investment, which allowed them to move to more productive work. Suppose you were one of these peasants. Frankly, you were not thrilled with farming. You worked hard and you did satisfactorily, but you could do much more. You worked well with leather, and could make harnesses better than anyone else on the manor. You made a deal with the other peasants, whereby you would specialize in harness making and leather working generally, and traded your superior work in that area for their superior corn or barley. Now for tanning the leather, it helped to have access to a river or at least a stream, so you moved your hut to the local waterway. Pretty soon another peasant who could do good ironwork set up a blacksmith shop near your tannery—the industrial city was begun. As peasants, you had been recognized by your peers and your baron (and ultimately your king) not just as labor motors, but as individuals having a specific new right, a right to a fixed percentage of the product of your labor.

Your baron, who was now wealthier than the other barons, was able to afford a larger military, among other things, and this proved beneficial later on, because as the people prospered, as little industrial cities become larger, word got around to the estates of other barons, and some of their peasants fled the land and moved to the city. By leaving their old lives of total servitude and destitution behind and moving, not so much to the new deal baron's manor but to the cities thereon, the loss of the newly freed peasants weakened their old barons. Remember, land and serfs combined equaled wealth to the old-timers. If a baron was losing peasants, he was losing wealth. What would you have done if you were one of these barons? The first idea was usually to mount up an expedition to the city, seize "your" peasants and bring them back by force. You would be disappointed. You would have to invade the land of the "industrial" baron, and he would have every incentive to keep the peasants who had come and the increased military power to do so. He would stop you, in the course of which he might even assert that the peasants in question had "sanctuary." If necessary, he would fight. He would, so to speak, spank your troops soundly and send you home. So what then? The baron losing serfs could do one of two things. He could decide that he would compete in offering economic advantages, letting his remaining peasants participate in the new economics and grow that way, or, if he was a really hardline reactionary, he might become an outlaw. As more estates came to accept the new ideas and cities began dotting the landscape, inevitably trade between them began and increased. The outlaw baron would be one who would raid the trader's caravans and demand tribute from them for crossing his land. If one goes to Europe today, particularly Germany, one might take a river cruise, and pass by some curious ruined structures by the water's edge. Upon asking you would be told that these were places where medieval lords ran chains across the river, and the barges of the merchants had to pay a toll before the lords would order the chains lowered and allow them to pass.[6]

It needs to be said here that this is not the only way the change from feudal life to city life occurs. There is what we may call the "colonial route,"[7] which is much less certain, less early in time, and usually takes longer to happen. It is often more of a detour than anything else. In the colonies established by states like the British, the Germans, the Spanish, the Portuguese, and others, notably in the Americas, Asia, and Africa, the ruling classes of the "mother country" set up feudal arrangements to reward their followers who helped conquer the areas. These leaders were given baronies or manors in these areas—in Latin America, they are often called *latifundias*. They were vast estates where the barons forced the natives to labor for them. Of course it did not happen exactly like this everywhere. In some places these large estates did not occur as planned and did not last or were abandoned. The early colonies in what became the United States, especially in the northern part, are examples of how the *latifundias* were defeated.[8]

Ultimately the colonies, already oppressed by their nonlocal rulers and arbitrarily divided into administrative regions for the benefit of faraway kings, revolted for independence. For instance, when Spain showed weakness against Napoleon, the local barons in South and Central America, who had been relatively independent before, revolted. In line with the common and necessary decentralization of the feudal state, they now chose full independence and control for themselves alone. Why pay a king who was doing nothing for them? This is pretty much in line with usual feudal understandings and the typical movement from centralization to decentralization to independence and back again. If we can take Bolivia as typical here, the country became independent in fact in 1824 after defeating the royalists in battle. Simon Bolivar became president and tried to make some basic changes to turn land over to the peasants. He failed, and one of his associates, Sucre, took over. He put the system of forced labor back in place (here, as in Mexico, it was known as the *encomenderos* system) to pay the bills incurred in the war for independence and keep the Hispanics in the high positions to which they had become accustomed. These Hispanics continued their almost uninterrupted rule and exploitation, supporting themselves by the labor of the Indians either by forced labor or by ensuring that it was the Indians who paid almost all the taxes. A succession of revolutionary movements came about, the strongest of which was the Nationalist Revolutionary Movement. The control of government vibrated back and forth between the leaders of this movement, who mostly just provided for themselves and paid off the Indians in exciting talk, and the military, which backed the Hispanics. Sooner or later, everyone except the peasants got to run the country. In 1980, even the old German Gestapo got into the action, with Klaus Barbie recruiting mercenaries to try his hand at rule.

Even when things were more settled—quite recently—problems continued. Case in point is that in 2000, an international corporation, Bechtel, began running the waterworks in Cochabamba, a city high in the Bolivian Andes. The area is arid and people are poor. The country as a whole is as poor as one might expect, with a handful of people, mostly descendants of the old Hispanics, getting almost all the wealth of the state. It is still highly agricultural, with a few resources such as natural gas that are diverted elsewhere.

Bechtel, which is not known as a corporation tenderly concerned with goodness and populism (and no more delicately considerate a company would move into this dangerous situation anyway), mindful of the problems with the seizure of property there in the recent past, limited their investment and said that people would have to pay more for their water to finance the company's replacement of outdated equipment, such as eighty-year-old holding tanks. Moreover, since people in the south of the area had no water service, prices would have to be raised to cover the costs of extending the service to them. This angered the peasants, who had no idea of what capital investment was or how it worked and no conception of paying for anything, having been kept in ignorance by their rulers. The result? "After days of protests and martial law, Bechtel . . . was forced out. As its executives fled the city, protest leaders pledged to improve service."[9] And so the previous local company, Semapa, run by the politicians, took over again. Years later, the southern area still had no water or sewerage service, and since the price of water had been cut, the company had no money to finance any extension. For those with service, they got it irregularly, in some places only two or three hours each day. It would take about five hundred million dollars to supply water to all, and while people talked about foreign aid, there was no aid. Who would loan them the money? Even bankers are not that dumb. Loan the money, and most of it would wind up in the pockets of politicians. Bankers have little use for corrupt politicians in such a small state, where the returns are not grand enough for the risk taken. Even if corruption was ended, what assurances did anyone have that the Bolivians would know how to keep the equipment running, and that they would start saving money for replacements as needed? It is evident that nothing would be accomplished. People need to do things themselves so they learn and have a stake in what happens.[10]

What can be done in these cases? One must get lucky, and I suppose luck was often a big component when we saw how some manor in Europe got a baron who started a new deal. But in colonial cases as Bolivia's, the matter must be handled by the national politicians; you need the luck of finding one who is sufficiently unusual to do the useful thing. In Bolivia's case, they apparently have lucked out in their new president, Evo Morales. The details are these: In late 2005, after more than fifty years of what was called "land reform," one hundred families in Bolivia still owned twenty-five million hectares of farmland, passed down from colonial days, while two million peasants owned only five million hectares.[11] Peasants continued to hope while they starved and rioted. Other reports were equally gloomy: one apparently done by the local Catholic churches found that those one hundred families owned 90 percent of the land overall.[12] In June 2006, President Morales, as a start, handed out "roughly 9,600 square miles of state-owned land to poor Indians."[13] Most went to individuals; a little went to tribes. The land is good land in the lowlands of the east, Bolivia's most fertile area. Some of the land was being used by the wealthy; some was unused. None was taken from the large landholders, although the plan is to "eventually seize and redistribute privately owned land that is unproductive, was obtained illegally or is being used for speculation."[14] The big landholders and the businesses that are simply adjuncts of the state (that is, that behave as crony capitalists) have objected and threatened violence if the land is taken from them, which is interesting since their ancestors obtained it by violence, and it has been held right down to the present because of that violence.

It is also interesting to note that the peasants have higher productivity and supply more food to the area than the *latifundias*, which are usually single-crop estates that produce export crops like sugar, and which are "dependent on the usage of large quantities of pesticides and fertilizers."[15] It appears as if Bolivia may be positioned to make a shift from politics to economics, from the state structure to the development for popular benefit for everyone, but led by the people in the cities. However, this is just a first step to start the peasantry on the path to developing their own cities as is done in other states without the colonial burden, or at least with that burden far behind them. It is a development that is centuries overdue, and at least in Bolivia's case, seems to be happening in a situation ripe for violence. Much of the violence is typically directed against the people who seem different, the outsiders. As a case in point, many Japanese people settled in Bolivia after the Second World War. They worked hard and bought land, land that had originally been stolen from one or another tribe. Now, with a government committed to land change, many of these people, innocent purchasers, are under fire. One of them recently remarked, "Now I get death threats from the landless peasants, and they are threatening to kidnap my family."[16] Many of these people will be killed as the price of development. Is this hard and cruel? Drive people out of the normal channels of development, retard their growth for centuries, and you get hard and cruel. No one has yet figured out a soft and gentle method of delivering people out of tribalism and into city life.

Bolivia does have at least one industrial city forming, El Alto, high in the mountains above the old political town of La Paz. While it is just in the early stages of development, many signs are good: women are emancipated from much of the male chauvinism that is so damaging in Latin America (many have jobs and they lead the family), the money economy is important for the first time to most of the inhabitants, and the city is growing in economic and political power. Most important is that people have hopes that they will live better and their children will live much better. For the first time in its history, Bolivia has the opportunity of ultimately obtaining solid, widely spread middle-class values and a degree of capitalism. They can now play catch-up to the more advanced areas of the world.[17]

Back in the usual noncolonial case, we were at the point where merchants from cities were running caravans of goods from one city to another, only to face banditry from some of the local barons and lords, that helped to join the city people to the king, at least temporarily.

The lawlessness of these latter kinds of local outlaw barons in many places provided the opportunities for kings to add to their own wealth, power, and glory. Kings probably felt they needed some kind of comeback. It seemed to them that the decentralizing nature of feudalism had all too often left them with fancy titles but not much power. With the advent of merchanting, they found opportunities. One area in which kings could gain some power or leverage was coinage. With developing trade, a system of money became useful. It was not uncommon in seafaring states to have money, but often they had a variety of money from many areas; it was new for landed states. In many places, kings took over a monopoly of the issuance of money, usually coins of gold or silver.[18] They frequently got to have their bust placed on the coins, which was a pleasing, ego-boosting thing for them. And the kings made money, literally, by establishing a royal mint: for a percentage of the gold or silver brought to them, the mint weighed it and gave it back to the merchant in precise amounts of coins, less the king's take for minting. The king was now in business, although no doubt it was quite impolitic to put it that way at the time.

Moreover, the traders in the new cities came to the king with a problem: some of his barons were raiding their caravans. These bandits were disrupting trade and therefore limiting the amount the king got by way of taxes. Could the king, the merchants asked, please do something about the bad guys? And, by the way, would the king please accept all the gifts the merchants brought to him (sorry, but one of his barons stole some of the presents from the caravan coming to see his majesty)? The king, no doubt, was properly outraged that some of his stuff had been stolen, and promised aid. When the caravan set out next time, some of the king's soldiers went with them for protection.

As you see, the complexities were mounting up: life was not as simple as it was in the good old days of feudalism. At least that was the way many barons thought about it when the king shut down their banditry business. But where did the king recruit his army, particularly the officers? Out of his noble class, that is, from the people who were the bandits. After all, the barons often knew nothing else but how to fight. The least competent of them were unable to run an estate with cities on it, and they were reduced to penury after the king stopped them from robbery. Their only skills were bashing other people over the head and leading a few cronies in the same line of work. They were perfect officers for the time. Over a few generations, these officers were weeded out by the kings and replaced by officers with a bit more military professionalism. This newer breed of soldier was offered ribbons and medals, kingly approbation, and a decent salary. They were not inspired so much by the robbery and rapine of their thuggish ancestors. Clausewitz, the foremost early authority on war, says regarding motive, "Of all the noble feelings which fill the human heart in the exciting tumult of battle, none, we must admit, are so powerful and constant as the soul's thirst for honour and renown."[19] Clausewitz was an officer who knew the pull of these feelings firsthand.

The merchants had other reasons to appreciate the kings over the local nobility. Our Baron Pushgas, now the thirty-third in line and much mellowed over his ancestors, still had the class attitudes of his ancestors, if not as much power and vigor to pursue them. But the baron maintained his castle and the lands and peasants, although his influence was greatly diminished. He was still the baron, even if he did wear third-hand armor plate and his castle was roasting in the summer and freezing in the winter, far more so than the smaller and nicer homes of many merchants. Even if he was poor by merchant standards, he had his servant put up a sign: Tradesmen Use Rear Entrance. If you were a wealthy merchant, you were still, he believed, his inferior, and he would certainly never consider marrying your daughter, even if she was attractive and brought a huge dowry with her. Our baron was a class snob who looked down on the tradesman, all the while envying him his riches.

So if you were that tradesman, it was an insult (yes, you are a bit of snob now too, looking down on your baron as an amiable idiot, not smart enough or rich enough to be interesting) to have to deal with this dolt. Far better to go see the king, who looked down on everyone equally.

The king's court was becoming more powerful.[20] Many of the old nobility had become subservient to the monarch—after all, he paid their salary as officers of the court or soldiers. By the way, note how very similar this is to China in the time of the Ming Dynasty, as we have discussed. There, as in Europe or anywhere, when the old nobility became part of the court, they lost their independence but gained influence over the crown. Other things happened to barons: landless nobles went into the military, or, as in Europe, into the official church as well, which was a special haven for the dispossessed second or third sons. Kings now had money and military muscle. They developed a civilian bureaucracy of baronial offspring as well, to keep track of everything for them. In time as the kings grew again in power, they no longer found it quite so convenient to be accessible to the merchants. The kings found the old nobility now bound to them. These people were more manageable; the kings were more comfortable in dealing with them than the developing capitalist and worker classes—which had a great deal in common at this point, for the laborers used to be peasants, as did the capitalists, and with a bit more saving and diligence, the workers might aspire to become capitalists themselves. Ultimately the monarchs switched their support and turned on their former allies.

How awkward! This was a time when political and economic power in the form of classes were struggling for preeminence; economics had emerged as the great energizer of society, but the old political system of kings and nobles still commanded the heights of power. In many ways, in different places, the old system of exploitation still retained the substantial ability to prevent capitalism from developing more rapidly and extensively. For example, in Christian and Moslem states alike, where religion was powerful and where the mullahs and priests often came from the old nobility,[21] there were barriers to usury, which was regarded as against God's law. In all practicality, this meant merchants and the city folk, all former serfs, were generally arrayed against the interests of the political, ruling class. Borrowing money at interest (which is what usury is), was typically the way people moved from peasants to laborers to capitalists. If you were just off the farm, you had no way to advance unless someone gave you help. If you were not related to someone already in the city, you needed to borrow. No doubt this concept of usury, as a means for cooperation in the newly complex society, must seem strange to many, even today. In fact it was a means for bankers, for starting or even established capitalists, and for starting workers dependent upon jobs, to work together. So if you wish to reverse the process of development and turn capitalists and laborers back into serfs, ban all loaning of money and kill the usurers. The old land-owning, serf-owning nobility will be back in total control again.

As a matter of course as a capitalist, you broke the law against usury. With luck you were not caught, but the penalties for borrower and loaner alike were dangerous. So if you were a capitalist, or if you had hopes of becoming one, what recourse did you have? You needed some control over the political system. You were part of what was now, or what was becoming, the middle class, between the nobles at the top and the peasants on the bottom, and you needed tools to oppose your interests to those of the nobility and even the king, who had proven to be untrustworthy. You would have to work as a class for political freedoms,[22] for the freedom to organize and apply influence to the state, to have a voice and a vote, to have the chance to elect a parliament or congress subservient to your interests in economic growth. For the first time in human history, a class developed that at least partially recognized that its own interest lay in ending exploitation, not simply adapting it for its own uses. Who needed to steal from others when it was more advantageous to produce? Who needed to enslave people to one's will when by cooperating one could produce much more for all who participated?[23]

As this kind of state, more and more city-led, developed in a way pleasing to the city people, it became more and more what Oppenheimer called the "modern constitutional state."[24] We have two developmental tracks now, the political and the economic. For the most part we need to look at them separately, although they intertwine and weave back and forth like the warp and woof of a sometimes very poorly woven garment. Complexity breeds problems in understanding every bit as much as it creates opportunities for growth and development, and the difficulty of people undergoing this development is quite well matched by the difficulty of people looking back and trying to understand it. We will be helped by the fact that in many places today, city development ranges from the "not yet happened" to the "just starting out" to the "bouncing along just fine, thanks," to the "hey, look what happened to us, isn't it wonderful" parts.

The example from England of the Magna Carta was, as shown earlier, a good illustration of how the transition to the city can be eased, as the barons had to give the merchants some benefits to aid them in their efforts against the king. The point about the king having to pay for the property he wanted is a case in point,[25] and we find other instances as well—for instance, a promise in that document from the king decreed that not only London, but "all other cities, boroughs, villages and ports shall have all their liberties and free customs."[26] Various procedural matters were also covered, such as limiting the king's courts as to how much they might charge certain classes of people for various offenses. On the subject of property, one provision was (again, as a king's promise to his subjects), that the king could not take the trees of anyone against the will of the person that owned them, a key point for the barons.[27] And finally, one might point to the protections of law for all, that, for example, "No free person shall be taken captive or imprisoned or disseized or be used badly by the law or be exiled or in any way be destroyed, nor shall we go or send [others] against him, unless by the legal judgment of his peers or by the law of the land."[28] One may recognize some of the language as still in force and effective today.

In places where that kind of cooperation between the barons and the rising classes was not available, other measures prevailed, and the fighting was often intense—as in France, with the use of "Madame Guillotine" to eliminate the recalcitrant nobles. If one would look to British colonial possessions in both the present United States and India, revolt was needed. To this day, India is still undergoing change in the development of areas into cities, with blighted rural areas tagging desperately along behind, still trying to shed extraordinarily corrupt politicians, caste prejudices, and failed or nonexistent educational opportunities. India has a rising Maoist movement in some impoverished rural areas, particularly in the state of Chhattisgarh, where villagers are forced to attend guerrilla rallies and watch the guerrillas dance and sing.[29] It is a form of entertainment where not attending might be visited with even more pain than would occur for one to have to listen to and watch the performances. People in this obviously reactionary movement may admit that Maoism in China did not work. They do not see the improvements around them in India and elsewhere that render Maoism obsolete. They talk about the values of the local indigenous peoples, or *adivasis*, as they force these people to support the nonproducing guerrillas out of the starvation-level standard of living the peasants now have.[30] These guerrillas provide an obvious example of exploitation of the poorest elements of society. It should surprise no one that, for example, one of their Maoist leaders, a "Comrade Kosa," gave up his technical schooling and has not worked, or further educated himself, since he learned how to cheat by living on the earnings of others without their consent.[31] The revolution in India continues for higher standards of living and for the gradual replacement of autocracy with democracy and a stagnant, feudal economy with a dynamic one. But it is hampered by widespread corruption in politics, and most of all by the general attitudes of people like the Maoists and other politicians who are unwilling to change, or do so only grudgingly. If one were to pick one thing that holds back developing areas such as India, I think I would reasonably say it is the attitudes or values that people hold. The values are changing, but it is a slow and quite frequently painful process. It is hard to keep up with evolutionary necessities, although not as hard as it would be in the long run to try and avoid them.

So we see another factor entering in: it is not just matters of economics and politics we must discuss (the economic means and the political means, in Oppenheimer's view), but we find these are tied to the values people possess, and by the way that these change under evolutionary pressure.

Suddenly we have four items we need to factor in—politics, economics, evolution, and values—each of which changes in different situations and presents itself in different ways. Perhaps we can start by saying that in the beginning with simple states, we moved to politics as the controller and determiner of all; and the values of feudal society, as I discussed in chapter 2, reflected the controls of the exploiters, just as those who exploited were themselves determined by the values of the society. The simple ways of hunting and gathering or farming, and the values those people had, tribe-specific loyalty for example, were being displaced. But as we leave tribalism, feudalism, and the maritime state behind, we create and are created by new values and a new economy: a city, or dynamic, growing economy, with the values needed to establish and maintain itself. To state the complexity here and only very partially untangle it, values (ethics) can create either a political state or an economic society, or a mixture of these (which is what the industrial city is), and the driving force behind all this is survival and change as a species, which is to say, evolution, or basic natural processes, which affect people quite as much as birds or cats, snails or daffodils.

But here we are concerned with the city, and so let me pick out the first strand in the tangled skein and talk about economics, which will mean that we need to talk about capitalist systems.

Following Oppenheimer, we can point out that people act to achieve what they believe will better their condition; they may act using a political system, that is, by appropriating the goods, services, and/or values of others, or they may act using economic means, that is, they may produce, or produce and then trade, for the goods, services, or values that they believe they want or need.[32] Those are the only two options that people have ever discovered. This does not mean that there may not be others awaiting us at some point in our evolutionary development.

Just as the politician (gang leader, dictator, president, king, and so on) uses force to make others turn over their goods, so the person using economics uses an exchange of values to get what he or she wants. And to produce, one customarily finds a need to have some kind of accumulation. We call this *capital.* When the politician uses force to get goods, and when this is the only (or at any rate, the dominant) method used to satisfy the desires of the ruling class, we do not think of it as a capitalist system, but we instead use the awkward phase, a *command economy*, to describe it.[33]

If we were to simply discuss the economic system of any society in which no political mechanism were present, or if it was extremely small and seldom used, we would be talking about a system of what is called *laissez-faire economics*. That is a system in which there is no force or fraud applied, and government has no role to play, or only the small role of preventing some people from using force or fraud against others, of insuring the peacefulness of the system. Some supporters of laissez-faire might even go so far as to argue that government should not use force for taxing people to provide for its activities, as Ayn Rand and others have suggested,[34] but such a very limited state, taxes or no, is something rather different than anything that has existed, although later on in this book we will find the concept reemerging. So while we may look at laissez-faire as an ideal type of human action,[35] thus far in human history such a system has not existed, however much some people may have boasted or complained about it. Since this kind of economics requires substantial freely chosen cooperation with others for mutual benefit, and is thus more complex, it comes much later in development than types requiring a great deal of force and therefore exploitation.

What we can do, rather than engaging in this kind of speculation, is simply point out that in the industrial city and in our world to date, we have an intermingling of economics and politics. We have a wide variety of arrangements, ranging from situations where politics has only a modest domination over economics, to systems where the two are more closely combined, as in mixed economies or planned economies. This last type is sometimes today referred to as *crony capitalism*, a term of derision used to point out the corruption involved in such a system, as, for example, when the Halliburton Company or Bechtel Corporation have "friends in high places" and use their political connections to obtain a monopoly position in a nominally independent economy, or where they simply get government contracts without having to show they have useful goods or services to offer.[36] The people who say they oppose this variety of capitalism usually protest that is not the way the state is supposed to work, but in fact that is the kind of thing states have always done. We can go at least as far back as the Greek philosopher Plato, who in his dialogue *The Republic* has the character Polemarchus argue that "justice consists in helping one's friends and harming one's enemies."[37] Precisely so. No one who "does business," as they call it, in any of the world's capitals would fail to understand that, nor, having seen how feudalism and the early city work, should we. He or she might be indignant about missing out on a part of the spoils or might miss them altogether and complain about his or her lack of friends, and then set about to make a "better" set of friends. That is a way a great deal of the business of the world is done under the mixed systems. There are people who cry "No! Favored people are not supposed to benefit, the benefits must go to the masses!" If they believe that a large state, which hands out favors and benefits somehow will give these only or even primarily to the least powerful in the system, they are pretty slow about understanding the system they support.[38]

Another term used to label or describe this variety of mixed system is *monopoly capitalism*. The term suffers from an unfortunate association with Marxist thought,[39] but if we strip away that association, it may be taken to mean simply that the capitalists with connections in the state desire a monopoly on the sale of the good or service they produce, understanding that they can then sell their stuff at a price much higher than an unencumbered market might allow. It may well be that this kind of arrangement is the ideal of mixed economy people, and so the term is not without its uses.

A final, if quite dubious type of capitalism is *state capitalism*. In its simplest form, this refers to a system like fascism, where all businesses are put directly under the control of the government in every detail: the state decides how much a business may charge for its product, what it may pay its workers, and even what, if anything, the business may produce. Businesspeople and workers alike are organized into groups called "syndicates" so they may more efficiently be made subject to state control.[40] It is not capitalism at all, but a reversion to feudalism of particularly strong type, without the fatally endemic decentralization of typical feudalism. It is, as Mussolini, the most important fascist of them all, pointed out, "totalitarian," because "for the Fascist, everything is in the state, and nothing human or spiritual exists, much less has value, outside the state."[41] The alleged alternative to fascism is *state communism*, where even nominal ownership is abolished, and all enterprises are part of the state apparatus. Both types are quite efficient at controlling people for the benefit of the rulers, but neither seems to be particularly capable at producing goods and services for the benefit of those outside the ruling class and hence are not viable in an evolutionary sense, as they work only for the exploiters/cheaters.

It must be evident from this that it is the mixed form we are interested in here, which, at least in its early stages in most cities, drove the development and changes of all sorts—aesthetic, religious, social—by an ever-changing combination of economics and politics. This is the type that Jane Jacobs talked about as providing the development necessary for ending feudalism.

How did development occur in cities? We have touched upon its beginnings, but more needs to be added by pointing out that as cities started, people specialized, and better goods were produced. Soon the baron's wife, for example, who had been insisting that the lovely tin pot she saw being carried about by an itinerant peddler was just what she wanted, persuaded the baron to get a local craftsman to make one like it; and the process of producing, copying, and expanding continued. More people wanted a pot just like that, then a craftsman put a different design on a new pot, which proved suddenly popular, and things changed again. Says Jacobs, "Economic life develops by grace of innovating; it expands by grace of import-replacing."[42] One little city replaced imports from another, then that other did something a little different or better—it innovated, and this spurred yet more growth on the part of all. Go from Venice to Singapore and wherever else you like, cities developed, and the "supply regions" around them,[43] while they might have become temporarily wealthy by narrowly specializing in one or two goods, ultimately lost out because they did not innovate. In time, as their goods became unnecessary, or produced better or more cheaply elsewhere, they returned to poverty, having never learned their lesson. Why is Cuba poor today, for example? Because under a dictator who could not think better than all "his" people in their own individual ways, that little state remains an agricultural supplier of sugar, which many other places can supply as well or better. What will happen to those supply regions like Saudi Arabia or Iran when other, less expensive power sources than oil are discovered? All their supposed wealth will disappear. There are no industrial cities in those states.

We also need to talk about democracy, which is evidently tied into the possession of a semi-independent economic sphere of activities. I have already pointed this out, but what needs developing here is a discussion about what is required for a state to be a democratic one.

Henry Mayo has proposed that there are four principles of democracy:[44] the people, the mass, must have some substantial control over those who make policy; the people must be politically equal; the people as voters must have an effective choice between serious alternatives presented to them on the ballot; and last, the people must be ruled by majority vote. These principles sometimes clash with each other, but not as often as they work together, and so any democracy is always a work in progress, with no democracy exactly like any other. Different people and different times work out the clash and harmony of principles differently. Note, by the way, that democracy is still, as with the majority rule principle, a primarily collective system. But as we shall see shortly, some individualist additions that allow some personal freedoms and protections of economic opportunities are present. In the city, democracy still usually trumps individualism. Nonetheless, without these principles being in place and being substantially effective, there is no democracy. By way of contrast, one may wish to look at the analysis of Fareed Zakaria, who suggests that there are two kinds of democracy: one is "constitutional liberalism" based on the kind of values like tolerance that we are about to talk about, and "illiberal democracy," which is simply more or less fair elections.[45] Let's briefly look at the principles, noting all the while the enormous complexity of having such principles over such simple systems as monarchy, the supposed democracy of Zakaria, or the commandments of a fascist ruler like Mussolini, who proposed as one of his commandments, that "Mussolini is always right."[46]

First, people must control those who make the decisions for them. Note that this is not to say that people, en masse, rule, because they do not. That would be *direct democracy,* a rather different system. Rather, it is to say that the mass must have a share of power in the system, and it must effectively be a share that people get when they have elections. Elections usually produce decision makers, not decisions, although sometimes democracies have instances of popular decision, as in most states in the United States, where people have a referendum to pass or defeat proposed constitutional amendments. In Switzerland, people in that state and in the subdivisions called cantons have the power to propose and pass constitutional amendments and particular laws; in some cantons, a good deal of the decision making is done in directly democratic fashion, however unusual this is in the rest of the democratic world.

The systems by which people chose their decision makers, their members of parliament or congress, vary widely, but generally the arrangements fit into one of two types: representation of people by *geographical method* (the US uses this) or the *proportional method* (used widely in continental Europe). The geographical system, or as it is sometimes called, the "first past the post" system, works like this: Any area to be represented is cut into geographical districts, roughly equal in population. The people in each district then elect someone to represent their district. Whoever gets the most votes is elected. If, for example, seven people run with the support of seven different parties, the winner probably will not get a majority of the votes; in fact, he or she might get only 30 percent, yet the winning candidate does not need a majority because the winner only needs more votes than any other candidate, that is, to be "first past the post." Under this system, the tendency is for parties to coalesce—party A may make a deal in order to defeat the candidate of party B: A arranges with party C that C will run a candidate in this district, but that A will support them, and that C, in turn, will not run a candidate but will support A in another district. Over time, these parties will make their coalitions stronger and stronger until they formally merge. Thus in the end, two major parties will emerge, and in most elections, one party will get a majority, since getting a majority is now necessary to be "first past the post." The advantage of this system is that it provides greater power to the parties in this semimonopolist position. Of course, as John Calhoun points out, this may merely mean more power to the dominant clique in each of the parties,[47] a dubious blessing at most since it probably just gives control to a dominating minority of the party and hence just becomes a variation of minority rule. And those who might prefer the ideas and persons in a smaller party (we call them *third parties* in this country) are cut out of any power. They feel they must vote for one of the two major party candidates and so they do not pick a candidate who appeals to them, but the one who offends them the least. When voters do this, they say that they are choosing the "lesser of two evils." Few people stop to think that electing the least obnoxious candidate may not be the best idea, but that there can be more positive ways to handle affairs of state.

An interesting alternative to this is a system whereby all the multiple parties run candidates for a congressional or parliamentary post, but if no candidate gets a majority, then a second run-off election is held, and the top two vote-getting candidates run against each other. This is guaranteed to produce a majority vote for one of the candidates, and seems to many people to be more democratic, even if they do have to endure two elections and two sets of irritating election commercials. This way, people do have more leeway in voting for someone who appeals to them, and gives them hope that their candidate might get into the runoff and might even get elected. It also gives the voter the option, having expressed his or her views in the first election, of skipping the second one. A person has had a choice and a chance to seriously participate and also a choice to not vote for the lesser evil. The French, for instance, do this in their presidential elections.

The second general kind of election system is known as *proportional representation*. It provides for representation of people's parties in the legislature according to that party's percentage of the vote, that is, according to how popular its ideas and candidates are. It works like this: there is no division of the state into districts, but rather each state (or canton, if we were to discuss Switzerland, which has one of the variations of the basic system), gets a percentage of the parliament according to that state's population. If we had this system and if the system had two hundred legislators, or parliamentary seats, and if your state or other subdivision had 10 percent of the total population, your state would get 10 percent of the legislators—twenty seats. All the parties interested in doing so would run a slate of candidates for these offices. They might run up to twenty candidates in the state, although many would run fewer. If party A wins 30 percent of the vote, then they get 30 percent of the twenty seats open, or six; party B gets 20 percent, and they elect four; even if party D or E gets only 5 percent of the vote, they still get some of the power of government, because they elect one candidate. These candidates have been selected beforehand by the party, and the winners are usually those at the top of the slate, as ranked by the party. So if you have backed party E, you still chose a representative who has a share of power. You have not voted for the lesser of evils—you have chosen someone you really wanted. Your vote has counted in a real and practical way. Compare this to the situation of the voter in the simple geographical system who voted for the losing candidate who "only" received 49.5 percent. Those who vote for the losing candidate lose as well; they have no voice in government and no impact on this system. Logically they might as well have gone fishing. Is the first system better because it concentrates power, and, one assumes, makes one party responsible to the voters for their governance, or is the second system better because it provides more accurate and effective representation?

The second principle of democracy is *political equality*. It seems obviously to be right and just to us today in the United States and in many, perhaps most, other countries. It was not so obvious many years ago, when only white male property owners over twenty-one years old voted in most of the American states; clearly we have arrived at an expanded understanding as we change as a species. When we lived in tribes, there was no equality except for the males within a particular tribe, because we were all interchangeable parts. Was the serf the equal of his feudal master? Or the wife, who was the property of her husband?

Can we say even today that people have a "right" to vote? I think not. In 2007, we still argue if convicted felons should vote. Why would we dispute over this if everyone has a right to vote? In the United States, the Bill of Rights sets out a number of our rights, such as free speech and a free press, to be secure in our persons, and so on, that apply equally to everyone who happens to be here. Even illegal aliens, for example, can publish freely. No right to vote is explicitly provided, although the Fifteenth Amendment of the United States Constitution does say that it cannot be restricted on grounds of race. If voting is a right like free speech or a free press, why can't aliens do it? If you are twelve years old, can you vote? No? Well then, get a lawyer, because your right is being violated! Does this sound a bit extreme? If the franchise is your right, you own it, and if you own it, it is like the shirt you are wearing: you can sell it if you wish. Do you want to suggest that you can sell your vote? Do you have a right to sell your vote? If you respond to the twelve-year-old vote or the purchased vote with some reluctance, then you do not really regard it as a right, and indeed it is not. It is extended to some people by certain governments at certain levels of development, and has no universality about it. If you lived on a desert island by yourself, would you have a right to free speech? Yes, but a right to vote? Nope. Voting is a governmental procedure. It is extremely valuable, necessary for democracy, a crucial means to the public good in contemporary systems—but not an absolute in all cases and circumstances.

Given that not all can or should vote, can we provide some test to decide who is entitled and who is not? The test, I believe, is one of minimum competence. That is, if voting is designed to allow people influence over government in order to provide for themselves and give support to policies they believe right, we need some minimal assurance that people have the ability to figure out what they want. Hence the twelve-year-old or the elderly person in the last stages of Alzheimer's may properly be denied the vote. On the other hand, the criminal and/or the person who sells his or her vote may be denied the vote on the grounds that their criminal records or intentions are directed to damage others, and since voting is not a right, we may exercise restraint over such persons and deny them the right to participate in the state in this way. After all, political equality and its practices are not absolute, but must be balanced against other principles. However, people who have been excluded in the past, such as women, blacks, and so on, do not show any lack of competence, nor do they engage in activities based on their race or gender to justify any exclusion on grounds of their danger to the rest of us or to the system. Therefore, they should have the right to vote.

Various reasons can be put forward as to why most of us should vote. This does not mean that most *will* vote, nor does it mean that we should require all to vote, although certain democracies do make it a law. One reason for voting might be that since we contribute to the support of the state through taxes, for example, we should have some chance to determine who makes our decisions; or, since it is usually thought that people have a right to self-defense, they reasonably may exercise that right by voting against candidates for office who would pass laws that might damage them. And it has been suggested for generations that having widespread participation in the political process is more likely to bring about useful results than any alternative. But I assume that reasons for political equality and the consequent vote that we have are sufficiently evident today to need little discussion.[48]

The third principle is that in democracy we are entitled to an effective choice between real (not imaginary) choices or no choices at all in our voting. There are several requirements that have to be met to ensure that we have adequately established this principle. The first is that when we vote, we must have our ballots counted fairly, without force being applied to coerce our decision, and without fraud being used to miscount the ballots or rig the election. While fraud and coercion were rampant in the nineteenth and over half of the twentieth century, even in the United States, it is clear today, with a former US president, Jimmy Carter, going around the world with other experts to help ensure the validity of the democratic process in elections, that we have long since accepted that elections, as a central and essential part of the democratic process, must be conducted fairly. Some states of course (Egypt comes to mind here) do not bother with such niceties,[49] but I doubt in most democracies or attempted democracies (and remember, not all the world is democratic even today, as China regularly demonstrates) that this would be terribly controversial. That the secret ballot and the safety of the ballot before, during, and after voting are needed is also quite clear, although with electronic voting, even in the United States there are questions about the safety of the vote count, especially as one of the major firms making voting machines is apparently a partisan organization with some problems in ballot security.[50] But we do try to insist upon the fairness of elections. Even winning candidates and parties would not be happy if it came out after the election that they won through unfair and illegal tactics. Thus, a disputed election in the United States, such as the 2000 presidential race, provokes extraordinary debate and controversy, as has the 2006 election in Mexico, where one candidate and his supporters threatened to destabilize that new and fragile democracy over the close loss of their candidate in the election.

It also needs mention that crucial to our effective choice at the ballot box is substantially equal opportunity to run for office. While democratic states generally impose some kind of requirement to put one's name on the ballot, such as paying a small fee or getting a few signatures on a nominating petition,[51] just to insure that one is serious about running, it is evident that insisting upon tens of thousands of signatures (as a few states in the US, such as New York, do for statewide office) or charging substantial fees is more of an attempt to restrict freedom of choice than anything else, and as such, should be eliminated from any democratic system when it is found.

Another part of effective choice is the existence and proper working of political parties. It is not totally clear why parties are necessary, but it appears that they may work to channel voters' interests, to discuss policy before the voters, and, in some systems, to offer compromises in policies and to show the voters alternative governments. This latter is quite evident in England, where the parties out of power will offer their party leaders to be prime minister. Also they will show the voters alternative cabinets to the one in power, thus giving voters more information on which to make a decision at election time.

The point that voters need to be informed about their potential choices ties directly into my final element of what is necessary for effective choice. Democracy is the only system that has any presumption that the people should be informed about what is going on with the state. It is felt that the more informed voters are, the wiser the choices they will make and the better government they will have. The government will better reflect the will of the people, because the people have a more efficient and substantial understanding of what their interests are and how they may be achieved under various policies and individuals. For this to happen, certain freedoms are needed.

The obvious freedoms are the freedoms of speech, press, organization, and assembly. If we say parties are necessary, then their candidates must be able to speak freely to the public and publish their positions, just as others must be free to comment upon them.[52] Democracy is a matter of the proper procedures as well as the underlying values to maintain those procedures, and here we consider these freedoms as procedural matters. If they do not exist, parties cannot function,[53] and if they do not function properly, effective choice does not exist. Note that these four freedoms are just what is minimally necessary for democracy; a substantial degree of freedom of religion seems to be useful as well, and since democracy arises out of capitalist class activity and a demand to be included (and finally dominate) the political process, it may be reasonably supposed that some amount of freedom to own and use one's property is needed as well. Of course capitalists as well as anyone else are perfectly capable of suppressing other people's use of property, as the use of eminent domain in this country to seize the land of the poor on behalf of large development corporations clearly demonstrates.[54] While exploitation is clearly hampered by democracy, it is certainly not eliminated. Still, it is clear that a person having property large or small, land or any other resource, can be more independent, in charge of his or her choices, than someone who is dependent upon others, including the state, for support. If your government gives you money for food, subsidizes your housing, and provides for you in sickness and old age, are you as ready to challenge those currently governing the state as a person who provides those benefits for himself or herself? It is unlikely, and this has been demonstrated, for instance, in Venezuela, where people who signed petitions to recall President Chavez are being fired (government employment is about 20 percent of total employment there), denied government benefits, and otherwise intimidated.[55] This is quite dangerous to democracy. Some independence of thought, occupation, and interest is necessary for democracy to work. That attribute of independence is a value we will be discussing shortly.

Our final principle is majority rule. We talk here about the majority of the elected members of the parliament or legislature making laws or otherwise binding decisions. As we have seen, these legislators, while elected by vote, may be chosen by a mere plurality and not a majority of citizens.

On what grounds do we assume here that majority rule is justifiable as opposed to some form of minority rule (unanimous rule will be discussed in the chapter on autonomy)? Minority rule has been commonplace throughout human history, with rule exercised by minorities of any kind whatsoever. Whoever could seize power did so.

One justification we can suggest is that when rules are made by a majority of legislators and when those persons are freely chosen by many citizens, then these rules are more likely to be obeyed. If they are more likely to be obeyed, then there will be less disruption of society occasioned by revolts or killing in the streets, and the real majority of us, who simply want to go about our business—making money, having fun, marrying and having families, and so on—can do it with only minimal disruption from the political sector. This is the same kind of justification one might have used in feudal times on behalf of the monarch's power against the continual tribal warfare of his time. The system lets us alone enough to allow us to live our lives in relative peace and security. It might be called a justification from expediency. The question might still be considered, however, as to why majority rule is more expedient now, whereas the rule of king and barons would be less so. The answer goes to the fact that we are changing as a species, and we are more open and accepting of the complexities of majority rule than our ancestors. Part of the answer also depends upon our greater acceptance of political equality, which is part of our change as a species, a movement away from tribalism where only the tribe counts. Since now we accept political equality, the notion seems to follow that if we each count equally, then a larger number must count for more than a smaller number; hence, it is appropriate that the larger rule the smaller. It is arithmetical equality. And last, we have what we might call the argument from despair: if the majority is incompetent, any kind of minority we might find is even more so. If, for example, sometimes democracies unjustifiably kill people, such as the United States carpet-bombing Germany and dropping atomic bombs on Japan to kill as many citizens as possible, still, the ethnic cleansing of the Holocaust in Germany and the murder of millions upon millions by the communist dictators like Lenin and Stalin in the Soviet Union, Mao in China, and Pol Pot in Cambodia were even more horrific. And supporters of majority rule also will quickly point out that democracies do not wage war upon each other, whereas other types of states do so routinely.[56] Of course, this point is open to question with the recent invasion of Lebanon, a weak democracy, by Israel, although that limited example may not be sufficient to disprove the general point.

We have the challenge in democracy to balance the four principles against one another. It is a continuous process. For example, if we have majority rule, may a majority eliminate the free speech of a minority in order to maintain their control? Most people who live in a democracy and want to maintain it would say no. This is an interference with free speech, which is a part of effective choice.

Strong democracies tend to set up what Mayo calls "no trespassing areas,"[57] which bar majorities from destroying the rights of minorities, because, if nothing else, that would mean that those minorities could never become a majority. Most of the first ten amendments to the US Constitution deal with just such concerns about the majority. For example, trial by jury is as much a process to avoid mobs deciding guilt or innocence as avoiding the simplicity, and opportunity for injustice, of the executive arresting, trying, and convicting those it wishes to punish. Federalism is a means of sorting out government powers to keep national majorities from infringing rights of minorities in particular states, by giving state legislatures some power: for example, supporters of a national constitutional amendment to ban gay marriage and civil unions, which are allowed in several states, have found that they cannot get enough support for their amendment. It is seen as changing the constitutional balance. And the Supreme Court has arisen as a defender of minority rights. Thus, even in matters allegedly involving national security, the Court has often upheld rights as it did in the "Pentagon Papers" case, where the Court said regarding the executive branch's claim to censor on the grounds of preserving security,

> To find that the President has "inherent power" to halt the publication of news . . . would wipe out the First Amendment and destroy the fundamental liberty and security of the very people the Government hopes to make "secure." No one can read the history of the adoption of the First Amendment without being convinced beyond any doubt that it was injunctions like those sought here that Madison and his collaborators intended to outlaw in this Nation for all time.[58]

There are two points to be made here. One is that such institutions are essential to protect freedoms. Other states may not use courts in this fashion, but some sort of institutional protections, and some sort of institutions that can be independent of other parts of government and not responsible to them, are needed. The other point is that institutions, courts or otherwise, are not enough. For the United States, it required people like James Madison and many others to press for freedom. If people do not support freedoms, freedoms will be lost, and as they are lost, democracy itself is lost. Freedom will never last "for all time" if the people reject it. People themselves, as well as the state, must accept the no-trespassing-zones concept. Presently there is a need to understand that in the long run, freedom is more secure than ever in places where it has at least a foothold. It is secure because it is needed for development. Evolutionary change is on this side of the issue. But the path to the future is not straightforward, and regression happens from time to time. It will be most unpleasant if this is one of those times in this country.

James Madison discusses how to preserve freedom and emphasizes institutional restraints. He begins by raising the question as to whether there are motives that might restrain the majority from damaging or destroying the rights of others. One possible restraint might be that people would understand that a "prudent regard to their own good,"[59] which is, after all, tied in with the long-range health of the community, might restrain them. Alas, people in the heat of the moment often fail to consider that long-range concern. And then could concern for the opinion of others restrain intolerant behavior? Well the "others" here are likely to be that very intolerant majority—one cannot be kept from the mob by the mob, and if one is to be restrained by the opinion of outsiders, remember that the views of strangers are not likely to be held of much account. Finally, religion provides no restraint, for members of legislatures are supposed to take an oath to look out for the common good, and an oath is about as solemn a religious statement as one can make; yet it has little effect, for individuals acting under the influence of the mob "often join without remorse in acts, against which their consciences would revolt if proposed to them . . . separately in their closets."[60] If one would provide control, one must rely more upon checks and balances that, by balancing the lusts and desires of some against those of others, provides a control over them all. A democracy without checks and balances cannot long endure, and is not worth much while it lasts.

Another of the American founding fathers, John Adams, discusses the matter in perhaps a more fundamental and profound way. I believe he has the deeper view of the matter. Having a balanced government is his great theme: he looks at the possibility of lodging all government power in one legislative body[61] and concludes that legislators, like anyone else or any other group, are too weak to be "trusted with unlimited power."[62] It is not that people, even politicians, are morally wicked, but they are too liable to be overcome by their passions or appetites. Says Adams,

> The passions are unlimited; nature has left them so. . . . They certainly increase by exercise, like the body. The love of gold grows faster than the heap of acquisition; the love of praise increases by every gratification, till it stings like an adder, and bites like a serpent. . . . Ambition strengthens at every advance, and at last takes possession of the whole soul so absolutely that a man sees nothing in the world of importance to others or himself but in his object.[63]

The three passions—greed, love of praise, and ambition—he believes overcome all others, overcome reason itself, and can so twist one's perspective so that one does not even know when they have conquered. Each time a passion is gratified, it becomes more demanding. As individuals, we must understand this,[64] as participants in a state system, we must guard against it, for people, he believes, "were intended by nature to live together in society, and this way to restrain one another, and in general they are a very good kind of creature; but they know each other's imbecility so well that they ought never to lead one another into temptation."[65] Always balance the powers and use checks and balances—but his hopes ultimately lie in convincing the citizens to support such a system. Without this support, without recognizing their own frailty, the desires of the many will quickly destroy the system on behalf of their own immediately perceived interests.

Education of people alone will never prevent the rule of whim, at least not at our present stage of development. Of course, we are continuing to develop; it may well soon be, and perhaps it is true now for some individuals, that they have evolved to the point where, as Daniel Goleman puts it, the passions cannot "hijack the brain,"[66] a point where some of us develop a balance or harmony that puts us in control of understanding what we do and why we do it.[67] Institutions are needed for checks and balances, yet without popular understanding and support for these ideas, the majority will destroy the rights of the minority and democracy will end. Balances, checks—these notions of complexity are woven, woof and warp, into the US political structure, the role of voters as well as senators, judges, and presidents. Democratic states may ignore the principles of the United States' system in particular institutions, but not the objectives, at the peril of losing democracy. And once democracy falls, the majority does not have control. A small minority, the innermost faction of the ruling party, comes to rule and complexity is lost for a time. Our evolutionary development can be set back, with all the attendant pain and suffering that disregard for reality entails.

Adams and Madison may have understood things quite well for their day, but it is still true today that democracy may fall victim to whim. What are the core values needed to sustain democracy, and are they strong enough to do the job? And do these values preserve the industrial city and the power of economics over politics? And finally, are these values good enough, and sturdy enough, that they will help us or our descendants to achieve further progress? Most obviously, what *are* these core values, beyond the evident ones in the complexities of economy and state?

In a general way, John Stuart Mill has shown these values to us—two fundamental values we must hold to have a city, a democracy, a partially free economic system, and to have development beyond these. These values can be summed up in the idea that people must accept independence and thereby responsibility for their own lives. Mill's position is that people need to understand that they are responsible for protecting themselves, and that they must become "self-dependent, relying on what they themselves can do, either separately or in concert, rather than on what others do for them."[68] For people to believe in protecting themselves, in depending on their own efforts, and relying on what they can do with others in a free, voluntary manner, requires a great deal of self-esteem and self-control. The cringing peasant, who sees his or her life at the mercy of the weather and the baron; the baron, who is even more cruel to his peasants than the king is to him; as well as the king, who acts in irresponsible fashion, making war instead of seeing to the crops that sustain his life, are none of them self-reliant. They obey, they issue orders or follow them, or a mixture of both, with little thought. Whatever the lord orders is right: "I was just following orders" is a feudal defense of one's actions that demonstrates that the person using it lacks any understanding of the idea of control over one's life. I suspect that the horror with which most of us view the actions of the Marines who apparently went on a killing spree in Iraq, killing innocent citizens with no cause nor even a pretext of cause[69] and then seeing their actions hidden by officials who were "responsible" for them, comes from the opposing values of most people today, who believe they are responsible for their own actions, and see the Marines as having "snapped," having lost control of themselves. In the past, not having control of oneself was routine, commonplace, accepted. The king himself was scarcely responsible, as was the baron—maybe it was the gods, the demons, or the weather. The individual committing the act was believed to have no responsibility for his or her own actions.

But in the city, the democracy, one minded one's own business, because that was what one did; and it was up to the other person to mind his or her own business. After all, if you didn't take care of yourself, who else should do it? And who else would know you like you do? No one else, it was felt, can handle your affairs better than you can. The more you do for yourself, the more you believe you can do. Actually admitting responsibility and acting up to that responsibility is the best builder of self-esteem, perhaps the only thing that does it properly. A "self-esteem" based on the praise or sycophancy of others is fake, compared to the real thing that is built on a foundation of actual accomplishment.[70] When one has self-esteem, one can help others without regret. Help can damage neither one's self-esteem nor that of others.

More specifically, what does this general value of responsibility entail? Eric Hoffer makes the point that it emphasizes work: city people are ready to work. They know it makes a difference in their lives and they like to do it.[71] In the past, Hoffer goes on, "work was viewed as a curse, a mark of bondage, or, at best, a necessary evil."[72] Indeed, the Old Testament of the Christian Bible says that God cursed man with work, that man should suffer "painful toil," not enjoy it, and eat the "thorns and thistles" that he had to grow in order to eat.[73] But today, Hoffer says, habits, morals, and needs have changed, so that we "should be willing to work day after day . . . and that work should be seen as a mark of uprightness and manly worth, is not only unparalleled in history but remains more or less incomprehensible to many people."[74] Labor in the days of feudalism, as I have pointed out, was for serfs or slaves, for inferiors. Hoffer believed the idea that labor is a sign of worth was not often found outside the Western nations, a belief that may well have been true fifty years ago when he wrote, but when one looks at the incredible development of the work ethic in India, China, Japan, Taiwan, Singapore, South Korea, and many non-Western lands, it seems to be sweeping the world.

Quite recently and delightfully, parts of Turkey seem to have joined in.[75] The change may have been waiting to happen for some time, and only now are things right to make it possible.

For the most part, the values of city people are closely related to work, self-protection, and self-reliance. Wilhelm Ropke provides us with a veritable checklist of these virtues, which are key to the survival and growth of the bourgeois society and include such things as responsibility for ownership, saving, personal effort, and "the courage to grapple on one's own with life and its uncertainties."[76] In the same way, yet with a more positive and enthusiastic twist, are the words of the English author and politician Auberon Herbert, who praises as "winning," doing "things for yourselves by your own exertions, through your own skill, your own courage, your friendly cooperation with another, your integrity in your common dealings, your unconquerable self-reliance and confidence in your own powers of doing."[77] Deirdre McCloskey has recently written a thorough development and explanation of the virtues of capitalism even beyond the wealth, health, and other extraordinary benefits it provides,[78] and Goleman, in his discussion of "emotional intelligence," is in fact talking about these same values.[79] If this analysis I provide here has good sense to it, such thoughts may become commonplace. I believe that underlying all these values is the great value of hope, the firm belief that what one does matters, that if one works one will be rewarded, that life will be better, and life will be even better for one's children. A state or society that attempts to deprive us of this hope is one that discourages work, that destroys initiative, and becomes one in which some people are regarded as mere means to the purposes of others, the others who cheat and exploit. This is fundamentally reactionary.

There are, of course, dissenters to this view and these values even today. When I was writing this chapter, I ran across a useful book review that was talking about a fundamentalist

Muslim sect, the Wahabis, and its impact on the world. The author of the review says that the beliefs of this sect have driven into "a common cause the world's wealthiest men, the sheikhs of Saudi Arabia, and the world's most oppressed children, orphans of the armed struggles in Afghanistan and elsewhere."[80] She goes on to point out the ties to terrorism, Al Qaeda, and so on, and indicates that perhaps the Wahabis are the most dangerous enemies of the modern world—which means, in our context, they are the enemies of the cities, the enemies of the development of economics and the improvement of life for all enemies of democracy and the freedom of peoples and the values that underlie all this growth. (Note that this is no reflection on ordinary Muslims.) It is not surprising to suppose this is so. This frightful alliance between the super-rich—that is, the rich who neither worked for nor deserved their wealth, who use it like the most reactionary feudal kings, to control their land and their serfs—and the extremely poor appears to be designed to throw the world back to feudalism, perhaps to tribalism. Of course they are not the only ones who have this in mind. There have been and continue to be alliances between the super-rich in the United States and extreme fundamentalist Christians, many of whom, especially in the South, are very poor. One need only think about an evangelist like Billy Sunday, for example (or the Pat Robertsons and Jerry Falwells of today), and his ties to the rich of his day, joined with his hatred of difference and development.

Or take Andrew Carnegie, who prospered mightily in the period of this country's great industrial growth, but turned against economics in favor of politics, talking about his belief that it was not more development that was needed, but rather to have the super-rich take control of everyone's life and goods and engage in "the proper administration of wealth"[81]—allegedly to benefit all, but in fact to allow the wealthy great power over all, that is, to put them in the position of the kings of old. Carnegie makes it quite clear when he talks about the benefits of state control of the prices companies charge, control exercised for the benefit of the owners of business, to prevent newer, better entrepreneurs to come along. He says, "It always comes back to me that government control, and that alone, will properly solve the problem [of prices]. . . . There is nothing alarming in this; capital is perfectly safe in the gas company, although it is under court control. So will all capital be, although under Government control."[82] So the alliance between the Muslim "street," the fundamentalist terrorist mullahs, and the Saudi princes operates today. It is no surprise to learn that Osama bin Laden is one of the princes and that almost all the terrorists who struck the United States in the 9/11 jihad are from Saudi Arabia, Wahabi's home.

What creates this truly evil alliance? Note here I say "evil." That is precisely what I mean. Herbert Spencer, perhaps the most prescient philosopher of the late nineteenth to the early twentieth century, had an excellent understanding of this aspect of evolution. He said, "All evil results from the non-adaptation of constitution to conditions,"[83] an idea that is echoed today, as Wilson comments that groups of people "who exhibit good traits are like to survive and reproduce better than any other kind of group."[84] Spencer goes on to point out that organisms are always being modified to meet the conditions under which they live. Hence, evil, he believes, is always diminishing, as we become more adapted to the needs of our environment. What Spencer does not adequately notice is that the environment itself is constantly changing—there is nothing static about it—and therefore more and more adjustments need to be made if the organism is to survive. As people, not squirrels or corn, we have developed a more and more complex society. As one example we've talked about, the baron, in order to gain more wealth, made his collection process more complex in order to do so, and the lives of the peasants were thereby made more complicated, challenging, and richer (in all respects—economic, social, moral, and developmental) as well. But the process of adaptation, indeed of constant adaptation to change, carries with it the need on the part of people to think about what they are doing, and thinking is hard business. When one fails to think, one is in pain, and it seems harder and harder to make change; it can even seem immoral to change, and that is the commonplace attitude of people who will not understand that the good old days were nothing of the sort. They are people like the old women who parade around the Kremlin on vanished Soviet holidays, holding up portraits of Lenin and Stalin, never mind that those leaders kept them in poverty and killed their children and grandchildren. They were accustomed to it, and that was good enough for them. To them, the new system, whatever that system may be, is immoral because it is not what is old.

Let's look at a recent and continuing controversy in this respect: same-sex marriage. The most dignified opposition to it comes from Senator Robert Byrd, who wants to make sure that marriage is totally defined by politics, but, of course, only by his brand of politics. He wants marriage kept the same as it "always" was. Those with different ideas and practices are not welcome. Says the senator, "The marriage bond as recognized in the Judeo-Christian tradition, as well as the legal codes of the world's most advanced societies, is the cornerstone on which the society itself depends for its moral and spiritual regeneration."[85] The author of the Web site that reports this speech introduces it by talking about "the Levite priest. He hadn't been in town no time and the sodomites wanted to rape him but they were given his concubine instead."[86] If one were to respond to this, it would be to say that marriage between a man and a woman only is past practice, it gives us no guide for the future; that perhaps people would be better off minding their own business (especially senators and authors of Web sites) than ordering other people to follow only one way of doing things. Charles Krauthammer, the ultimate American neoconservative, points with alarm to the threat to him that if you allow gay marriage, you allow any kind of marriage. People can do what they want, they are freed of state control, why they could even love more than one person at a time.[87] The most amusing response to his position was a blog posting that satirizes this position, taking the argument to its ultimate extreme:

> Equality is a slippery slope, people, and if you give it to the gays you have to give it to the polygamists and if you give it to the polygamists you have to give it to the serial dog molesters and if you give it to the serial dog molesters you have to give it to the machine fetishists and the next thing you know you're being tied up by a trio of polygamist lesbian powerbooks and you can't get out because the safety word is case sensitive![88]

There are not many cases as clear-cut as this. Do you favor economics and personal choice, or politics and state control of personal choice? Of course if you choose the latter, you not only stifle change, but you afford your masters—unless you have hopes that you will be a master and others your serfs—a chance to rule anything that their whimsy dictates, including ordering you to marry a certain person (as we saw with Kropotkin), not to marry at all, or to marry several people.

There is always a range of choices we can make: that is what free will is about, as more and more our evolutionary development does not straitjacket us. Make the wrong choice too many times, however, and we regress, we become less and less capable of adaptation as we saw earlier with the Chinese emperor and the admiral. I suspect that this is why the twentieth century saw so many massacres and brutalities. People were trying to simplify rather than deal with development, with economic and personal growth, and they invariably made the wrong choices. As a tribe, it was right to follow the leader. Today, that prescription leads to Hitler or Stalin, Mao or Amin, that is, to major disaster and evil.

There are many good signs for us, signs that some of us have pretty well understood why cities work, and have thus moved from cities to the *global city* idea, treating the whole world as a city. As some examples among many, there are people who have adapted by creating nongovernmental organizations such as Doctors Without Borders, which would have been unimaginable a century ago; we have the Geneva Conventions, which, however much they are under assault by the United States,[89] are being upheld elsewhere more than ever. Or, to take an example related to this, other countries (notably, European ones) are moving more and more strongly to defend the role of courts, rather than the use of the whims of rulers, to control terrorism. Spain, for example, is complaining that the United States is interfering with their judicial efforts to try terrorists in a fair fashion, in accordance with reasonable procedures and without torture.[90] And the United Nations has courts to try people accused of crimes against humanity, and will intervene, when it can, against such offenses as "ethnic cleansing," otherwise known as tribal warfare. Ask how many international court decisions were handed down before the second half of the twentieth century, as opposed to afterwards. Since no decisions exist in the earlier time, the answer is impressive. Are international courts all good? Of course not, but they are change for the better. They are more in accord with human needs now, rather than human needs during the tribal or feudal periods. And of course the development I mentioned earlier in Bolivia, while not leading immediately to development beyond the city, nonetheless is a positive step that will ultimately lead to the city, and then, in time, beyond.

Some of us—enough of us—are city people. And some of the city people are starting to make the change beyond that. Many of us have understood where we are going. Let me give you one story that I hope makes the point, at least in an economic way—it did so for me many years ago when I first heard it. It is story of the pencil, first written in December 1958 by Leonard Read. It seems simple on the surface, but there is a world of profundity underneath. For something written almost half a century ago in a small publication to find its way to the Internet today attests to the fact that more people than I have found it useful. It is called "I, Pencil: My Family Tree as Told to Leonard E. Read."[91] The essay is written to make the point that producing just a simple and ubiquitous thing as a pencil is so complicated that no one can understand how it is done, nor do it all by himself or herself. The story traces the steps of making a pencil, all the way from finding and cutting the trees or finding and preparing the ingredients of the "lead," that is, graphite, as well as all the other ingredients. For globalization, we have cedar from California or Oregon; graphite from Ceylon, which is mixed with clay from Mississippi, gets involved with hot candelilla wax from Mexico, and gets mixed with a bit of zinc and copper and nickel, which can come from a number of countries; and an eraser, which among other things includes rape seed oil from what was then the Dutch East Indies, and finally, but probably not completely, gets hooked up with pumice from Italy. All those things come together only by literally thousands of people cooperating—and this does not include hundreds of relationships, like people in the power company who supply the power for the people who blend the cadmium sulfide, and the guy who made the saw (and the people who mined the ore that was used in the making of the saw) that was used by the logger who cut the tree down. You have, Read says, "millions of tiny know-hows configurating . . . in response to human necessity and desire" in order to make this simple thing. Let it alone, says Read, let people work together, let people use all their "creative energies" freely for the benefit of everyone.[92] It was an important lesson for a teenager to learn, and I think that it is a basic lesson for all those who would advance us to the city, as well as beyond the city to the globe.

Of course becoming city people was a great advance for our species, even if a significant number of people have not made the change yet, either not yet progressing toward it, or just beginning to engage in the change. Change is there and it is happening. If we want to reduce the evil of our present world, we will work to adapt ourselves to our new ways of being. Reactionaries will say that we have no steadfast morals, that we are plastic, bendable as needs dictate. Indeed we are, if we would survive. For people understanding the change, they are indeed plastic. The reactionaries may make use of as much plastique in their bombs as they can, but the rigidity of these people is not going to defeat the new changes any more than they could permanently prevent all the other changes our species has undergone. We human beings are endlessly inventive; it is possible that someday we can be proud of our development.

NOTES

1. Jane Jacobs, *Cities and the Wealth of Nations* (New York: Random House, 1984), p. 232.

2. Franz Oppenheimer, *The State*, trans. John Gitterman, introduced by C. Hamilton, http://www.franz-oppenheimer.de (accessed June 14, 2007), pp. 88–96.

3. See the discussion in Jacobs, *Cities and the Wealth of Nations*, beginning on page 31. Note especially her comment on page 32 that "cities are unique in their abilities to shape and reshape the economies of other settlements, including those far removed from

them geographically." Or Oppenheimer, *The State*, p. 89, that "the industrial city is directly opposed to the state." (Emphasis removed.)

4. Henri Pirenne, *Medieval Cities*, trans. Frank D. Halsey (Garden City, NY: Doubleday, 1925), p. 19. Note also his comments on pp. 93–96 on the growth of the middle class and the rise of both industry and commerce in European cities. To Pirenne, as to Jacobs and Oppenheimer, it is the city that is the fount of economic growth.

5. Pirenne points this out clearly when he says "the love of gain—the desire to ameliorate one's condition must have carried . . . very little weight with a population accustomed to the traditional way of living." Ibid., p. 76.

6. See, for instance, comments on http://www.gayot.com/travel/citytrips/romantic_rhine/day1.html (accessed July 14, 2006).

7. Any competent text on Latin American history can talk about the colonial period, and most of them are capable of finding that exploitation existed. See, for example, John Charles Chasteen, *Born in Blood and Fire*, 2nd ed. (New York: Norton, 2006). So far I have not found one that is able to go beyond simple right vs. left categories, or can understand how this relates to exploitation elsewhere, and how it can be remedied.

8. Murray N. Rothbard, *Conceived in Liberty* (Chicago: Arlington House, 1975). In particular, volume 1 can be studied for evidence of this.

9. Juan Forero, "Who Will Bring Water to the Bolivian Poor?" *New York Times*, December 15, 2005, p. C1.

10. So how do people without water service get it? It is trucked in at very high cost, and as one customer said, "sometimes it contains tiny worms." And as the presumed leader of the people, Oscar Olivera, the man who led the move to throw Bechtel out, admitted, "we were not ready to build new alternatives." Ibid.

11. This doesn't include the quarter of a million peasants with no land, who farm for the large farmers, or who are crowded in what are called "cities" but which in fact have few if any city functions and are just large slums. See Lelia Lu, "The Agrarian Reform That Wasn't," *Upside Down World*, December 7, 2005, http://www.zmag

.org/content/showarticle.cfm?ItemID+9286 (accessed May 7, 2006).

12. Monica Machicao, "Bolivia's 'Agrarian Revolution' Sparks Hope and Fear," June 5, 2006, http://ca.today.reuters.com (accessed June 5, 2006).

13. Fiona Smith, "Bolivia Launches 'Agrarian Revolution,'" June 3, 2006, http://www.phillyburbs.com/pb-dyn/news/88-0603 2006-665571.html (accessed June 4, 2006).

14. Ibid. Note that motivation and belief have no roles here. In the article, Morales talks about this being a "nationalizing" of resources, when it fact it is a radical privatization that will lead to peasants moving into the middle class. A leader of the Guarani Indians told people that the "landowners, the foreign companies, the political parties" took the land from the people. No, it was the conquistadors—the current gang who run the rest of the country merely continued that exploitation. A local reactionary group, the "National Farming Confederation," was formed to provide "self-defense," that is, to use violence to maintain their system of exploitation. All this inability to understand what is happening is typical of change. One supposes that if people had the slightest comprehension of what they were doing, they might be afraid to do it.

15. Lu, "The Agrarian Reform That Wasn't."

16. Monte Reel, "Two Views of Justice Fuel Bolivian Land Battle," *Washington Post*, June 20, 2006, p. A1.

17. See Monte Reel, "Bolivia's Rural Women Are Remaking Cities, Lives," *Washington Post*, March 6, 2007, p. A1. As usual when people begin improving their lives, there are detractors. For instance, Linda Farthing, Juan Arbona, and Benjamin Kohl charge that "neoliberalism" (they are apparently unaware that this kind of development antedates the twentieth century) has caused "growing inequalities" in the area. This is the typical aristocratic hatred of growth of the "lower classes" into positions of economic and political power. See their "The Cities That Neo-Liberalism Built," http://www.harvardir.org/articles/1433/1/ (accessed March 14, 2007). A partial antidote to this sort of reactionary griping may be found in F. A. Hayek, ed., *Capitalism and the Historians* (Chicago: University of Chicago Press, 1954). An even better response is simply discovered

in the Reel article, where one sees the long overdue growth and development of people.

18. Perhaps the most famous of the coins minted by any state are the Pieces of Eight, minted by Spain starting in 1497. These one-ounce silver coins were so nicknamed because they were designed to be broken into eight bits, or pieces, to make change. Spain made so many of these coins, both in Spain as well as their colonial mints in Mexico and Peru, that they became the world's currency. For example, in the United States both before and after its war of independence, the Spanish coin was the favored currency, ultimately replaced by US silver dollars. Curiously, for twelve years in the nineteenth century, the US tried to imitate the Spanish success and put out a "trade dollar" as well.

19. Carl von Clausewitz, *On War*, ed. Anatol Rapoport, original edition published in 1832. (Baltimore: Penguin Books, 1968), p. 146.

20. Oppenheimer, *The State*, chap. 6.

21. Pirenne has an example from England, *Medieval Cities*, p. 125.

22. Pirenne says that to the "middle class was reserved the mission of spreading the idea of liberty far and wide," *Medieval Cities*, p. 154. It was a mission evidently in their interest.

23. See Pirenne, who points out that "the habit of cooperating is to be found everywhere in economic life," *Medieval Cities*, p. 84. See also David Sloan Wilson, *Evolution for Everyone* (New York: Delacorte Press, 2007), who talks a good deal about cooperation as evolutionary strategy. Curiously, he appears to accept exploitation as a kind of cooperation and does not distinguish between voluntary and forced conduct. See his comments in chapters 21 and 22 regarding the poor low-caste chimp named Jomeo, who "cooperated" by being forced to give up almost all his food to the dominant chimps who, I suppose, "cooperated" with Jomeo by letting him live to gather more food for them. A parallel might be drawn between the chimps and the barons and serfs in the feudal state. I am not sure who aped whom.

24. Oppenheimer, *The State*, chap. 6.

25. The king was not happy and tried to wiggle out of obeying it, even getting the pope to declare it null and void. He failed. And by the time of the final version of the Magna Carta in 1225, the English had solid and certain law. Even today, almost eight centuries later, a dozen of the original provisions, at least in part, remain law in England.

26. See Samuel E. Thorne et al., *The Great Charter* (New York: New American Library, 1966), chap. 13. My translation. Or online, for an English-only version, http://www.cs.indiana.edu/statecraft/magna-carta.html (accessed November 26, 2006).

27. Ibid., chap. 28.

28. Ibid., chap. 39.

29. Somini Sengupta, "In the Villages of India, Maoist Rebel Guerrillas Widen a 'People's War,'" *New York Times*, April 13, 2006, p. A12. Places like India—not colonized, just conquered—present a different picture from areas in Latin America. In India, Great Britain acted as a maritime state in exploiting people there. Elsewhere, the British Empire did colonize, as in the United States. But Spain, as our Bolivian example makes clear, was an occupying landed state.

30. Ibid.

31. John Lancaster, "India's Ragtag Band of Maoists Takes Root among Rural Poor," *Washington Post*, May 13, 2006, p. A1.

32. In Oppenheimer's words, "There are two fundamentally opposed means whereby man, requiring sustenance, is impelled to obtain the necessary means for satisfying his desires. These are work and robbery, one's own labor and the forcible appropriation of the labor of others," *The State*, p. 12. This is the "kleptocracy" of Jared Diamond, *Guns, Germs, and Steel* (New York: Norton, 2005), p. 276, and the cheating referred to by others.

33. See, for example, a comment on the "command" or "planned" economy at http://www.tiscali.co.uk/reference/encyclopaedia/ hutchinson/m0037989. The notion that the two are synonymous is confusing, since (a) a "command" economy is not an economy at all, but a total political system, and (b) a key difference between economics and politics is not overplanned versus unplanned, but between who does the planning—individ-

uals acting according to their own interests without the use of force, or individuals using force to achieve their own interest at the expense of others. (Site accessed February 17, 2007.)

34. Ayn Rand is one person who argues this point, and one may find some of the essays in her newsletter. See, for example, her comments in the "Intellectual Ammunition Department," *Objectivist Newsletter* no. 2 (February 1964): 7.

35. Ludwig von Mises, *Human Action*, 3rd rev. ed. (Chicago: Henry Regnery, 1966), especially the introduction and the first two chapters.

36. As one example, see Tim Shorrock, "Crony Capitalism Goes Tribal," *Nation*, April 1, 2002, http://www.thenation.com/doc/20020401/shorrock (accessed April 7, 2006).

37. Plato, *The Republic of Plato*, translated with introduction and notes by Francis MacDonald Cornford, 12th ed. (New York: Oxford University Press, 1955), p. 12.

38. A perfect example occurred in connection with Hurricane Katrina in New Orleans (actually, there are lots of good examples). People were left homeless, so the Federal Emergency Management Agency (FEMA) sprang into action. No more than a brief period elapsed, and FEMA had constructed trailer parks and put in trailers. We apparently don't know who they bought or leased the land for or how much they paid for the trailers, and who they bought them from. It is a reasonable assumption that they did not pay any of this money to homeless people. Estimates in one place, Morgan City, Louisiana, are that FEMA spent about seven and a half million dollars to build the trailer park there, which makes it a lot more expensive than any trailer park I have ever seen. In that park there were 198 trailers, of which 15 were lived in. FEMA officials refuse to tell anyone outside the government how much they have spent or why so few people live there. And they appear to have lots of money left, because they would let no resident talk to a reporter unless a "FEMA representative" and paid bureaucrat was there for the interview. Associated Press, "FEMA Muzzling La. Trailer-Park Residents," July 20, 2006.

39. See for a typical example, http://www.monthlyreview.org/1004pms2.htm (accessed June 1, 2007).

40. For those interested in more detail as to how this worked, I recommend G. Lowell Field, *The Syndical and Corporative Institutions of Italian Fascism* (New York: Columbia University Press, 1938).

41. Benito Mussolini, "The Doctrine of Fascism," first printed in the Italian Encyclopedia in 1932, later reprinted in many other places. My copy is from Carl Cohen, *Communism, Fascism and Democracy* (New York: Random House, 1963), p. 352.

42. Jacobs, *Cities and the Wealth of Nations*, p. 39.

43. Ibid., chap. 4.

44. Henry Mayo, *An Introduction to Democratic Theory* (New York: Oxford University Press, 1960).

45. See Fareed Zakaria, "The Rise of Illiberal Democracy," from the November 1997 issue of *Foreign Affairs*. It can be found at http://www.fareedzakaria.com/ARTICLES/other/democracy.html (accessed August 12, 2006). Note that he does not really accept that value-laden, bourgeois liberalism is required for democracy.

46. This commandment is the eighth in one version, tenth in another. See www.historyguide.org/europe/duce.html (accessed May 7, 2007).

47. John C. Calhoun, "Disquisition on Government," 1849, paragraph 64. The best way to find this, as with so much of interest, is to go online. This is at http://praxeology.net/JCC.DG.htm (accessed September 7, 2006).

48. Please note that this is not a defense of equality in general. We are not all equal, one to another, in every aspect of life. Rather, the argument here is simply that in a democratic system, we are all entitled to be considered politically and legally equal, in the same way that we believe today that justice ought to be blind, that is, no respecter of class or special privilege. We ought to be treated equally in any aspect of our relations to the democratic state in choosing a parliament or president or prime minister, in serving on a jury or being confronted by a prosecutor, judge, and jury. This is why the efforts of the current US administration (I write this in 2006) to deny people suspected—or allegedly suspected—of being terrorists any access to the courts, must be considered violative of basic democratic standards.

49. Among other articles, see Jeffrey Black, "Egyptian Trial of Pro-reform Judges Halted after Protests," at http://news.independent.co.uk/world/africa/article364319.ece (accessed June 1, 2007).

50. See http://www.blackboxvoting.org for some comments on this, or check out http://www.scoop.agonist.org/section_all_ for its section on "USA: E-Voting," a thorough compiling of articles on many aspects of voting problems and the difficulty with several companies on this. (Both accessed July 8, 2007.)

51. For instance, a few years ago, when I ran for city council in a small town in Kentucky, I needed three signatures to get on the ballot.

52. The Supreme Court in the United States has long recognized this point, and while the Court is unwilling to accept the words of the First Amendment at their face value (it says there shall be "no law . . . abridging" these freedoms, not that some laws are acceptable if a majority really want them), the Court has been insistent that the bar to restricting these freedoms in political contexts is higher than it is in other circumstances. See the "Pentagon Papers" case, *New York Times Company v. United States*, 403 U.S. 713 (1971).

53. For many years in Portugal under the control of the dictator Salazar and his political organization, the Falange Party, other political parties were technically allowed to exist, but they had no freedom to tell the voters they existed, what they stood for, or who their candidates for office were. These parties had no function in this system except perhaps a modest propaganda one. This way the dictator could claim he was fairly elected in a democratic fashion.

54. See the recent case in the United States: *Kelo v. New London*, 125 US 2655, decided June 23, 2005.

55. Helen Murphy, "Chavez's Blacklist of Venezuelan Opposition Intimidates Voters," June 19, 2006, http://www.bloomberg.com/apps/news (accessed March 15, 2007).

56. Of course these same supporters of democracy might find it hurtful to their case when one points to the invasion by the United States of Iraq and the subsequent bloodshed and atrocities in that state, some committed by US troops and mercenaries, as indicating the dubiously pacific attitudes of democracies. "Preventative war" is

not particularly attractive to peace-loving people, especially when they find that such a war is undertaken for "regime change" or some other reason, including, possibly, personal or family revenge. How many innocent citizens may one kill in order to change a form of government, to remove a naughty ruler, or to make one's president happy?

57. Henry Mayo, *An Introduction to Democratic Theory* (New York: Oxford University Press, 1960), chap. 9, especially pp. 183ff.

58. *New York Times Company v. United States*. Unfortunately, this case is not quoted so often in the present circumstances, where its points are extremely relevant.

59. In Saul Padover, ed., *The Complete Madison* (New York: Harper, 1953), p. 27.

60. Ibid., p. 28.

61. The idea of the "unitary theory of the presidency" that floats around today—essentially that the president can do what he wants without supervision by courts or Congress, that his opinion of what the constitution means is supreme—would be regarded by people like Adams as monarchist and not worth discussing. That point was settled by the American Revolution. In democracy we may not substitute "The president is always right" for our Constitution and Bill of Rights.

62. Adrienne Koch and William Peden, eds., *The Selected Writings of John and John Quincy Adams* (New York: Knopf, 1946), p. 98.

63. Ibid.

64. Adams himself fell victim to this. Defeated by his friend Jefferson in the presidential election of 1800, Adams felt cheated out of a second term; he turned on Jefferson, and egged on by his wife and the Federalist Party people because they saw themselves losing power, Adams signed the Alien and Sedition Acts, which were designed to destroy freedom of the press and put supporters of Jefferson in jail. The things he and Abigail were prepared to believe about Jefferson are amazing. The two ex-presidents, last survivors of the "founding father" group, were finally reunited in the last years of their lives after passion had faded.

65. Ibid., p. 99.

66. Daniel Goleman, *Emotional Intelligence* (New York: Bantam, 1995), p. 17.

67. Goleman suggests that the "complementarity" of various brain parts and functions, and that each element is a "full partner in mental life," lets us function well enough to the point where we may become somewhat more trustworthy than the kind of person Adams knew and was. Ibid., p. 28.

68. John Stuart Mill, *Considerations on Representative Government*, ed. Currin V. Shields (New York: Library of Liberal Arts, 1958), p. 208.

69. See early and brief accounts of what apparently happened, and how some officials tried to cover it over, in Eric Schmitt and David S. Cloud, "Military Inquiry Is Said to Oppose Account of Raid," *New York Times*, May 31, 2006, p. A1, and Richard A. Oppel Jr., "Iraqi Accuses U.S. of 'Daily' Attacks against Civilians," *New York Times*, June 2, 2006, p. A1. The attacks seem to have taken place in the town of Haditha in November 2005. By the time you are reading this, much more will be known.

70. An interesting science fiction novel built around these ideas is Eric Frank Russell, *The Great Explosion* (New York: Avon Books, 1962). Much of what Russell says is based on a fusion of the ideas of Gandhi and the notions of responsibility that were around at the time. His use of "MYOB," or "Mind Your Own Business," is illuminating.

71. Along this line, it sometimes seems that popular culture today emphasizes something we call a "slacker culture." Insofar as it does so, it is evidently reactionary, and it is rejected—once it is pointed out to people—overwhelmingly.

72. Eric Hoffer, *The Ordeal of Change* (New York: Harper & Row, 1952), p. 28.

73. Genesis 3:17, and "But woman got the curse of pain in childbirth and being ruled by her husband," 3:16. People may argue who suffered the most. But both were cursed.

74. Hoffer, *Ordeal of Change*.

75. Dan Bilefsky, "Turks Knock on Europe's Door, with Evidence That Islam and Capitalism Can Coexist," *New York Times*, August 27, 2006, p. A4.

76. Wilhelm Ropke, *A Humane Economy: The Social Framework of the Free Market* (Chicago: Henry Regnery, 1960), p. 98. Ropke is the economist who trained Ludwig Erhard, who as minister of finance in the West German government after World War II was crucial in establishing the policies and ideas that turned his country from a land of starving and defeated people into a prosperous, growth-oriented area. Ropke undermines his position by suggesting that these values are boring and dull; his defense of them is the back-hand one that "there is nothing shameful in the self-reliance and self-assertion of the individual taking care of himself and his family," p. 119. The development of these values constitutes one of the greatest advances in human history, and if this lukewarm support is the best he can offer to Germans, it is probably indicative, if this view is widespread in Germany, as to why those values are much in retreat in his nation today.

77. Auberon Herbert, *The Voluntaryist Creed* (London: Oxford University Press, 1908), p. 76.

78. Deirdre N. McCloskey, *The Bourgeois Virtues* (Chicago: University of Chicago Press, 2006).

79. Goleman, *Emotional Intelligence*, pp. 56–57.

80. Barbara Bamberger Scott, reviewing "The Wahabi Cult and the Hidden Roots of Modern Jihad," posted and http://agonist.org/20060605/gods_terrorists (accessed June 5, 2000).

81. Andrew Carnegie, *The Gospel of Wealth and Other Timely Essays* (Cambridge, MA: Harvard University Press. 1962), p. 14.

82. Quoting Carnegie in Gabriel Kolko, *The Triumph of Conservatism* (New York: Quadrangle Press, 1967), p. 18.

83. Herbert Spencer, *Social Statics* (New York: Robert Schalkenbach Foundation, 1970), p. 54.

84. Wilson, *Evolution for Everyone*, p. 31. See also right above this is the point where he emphases that "goodness can evolve."

85. Senator Robert Byrd, *Proceedings and Debates of the 104th Congress, Second Session* 142, no. 123 (September 10, 1996), as quoted on http://www.jesus-is-lord.com/senator.htm. The owner of the Web site is as responsible for his own grammar as he is for his ideas. (Accessed July 8, 2007.)

86. Ibid.

87. Charles Krauthammer, "Pandora and Polygamy," *Washington Post*, March 17, 2006, p. A19.

88. http://fafblog.blogspot.com, posted on June 4, 2006, and accessed June 5, 2006. (We take both sides of the blogging business here.)

89. See as one small example, Julian E. Barnes, "Army Manual to Skip Geneva Detainee Rule," *Los Angeles Times*, June 5, 2006.

90. Elaine Sciolino, "Spanish Judge Calls for Closing U.S. prison at Guantanamo," *New York Times*, June 4, 2006, p. 6.

91. Yes, we will ignore the terrible pun. It was first published in the December 1958 issue of the *Freeman*. I had to put aside my collection of the magazine when I moved some time back; I regret losing it, and am delighted to find this essay again, at http://www.econlob.org/org/ (accessed May 20, 2007).

92. Ibid.

CHAPTER 5

GLOBAL PEOPLE

Now in ecstasy we trace
The aspects of the human race:
Some are men and some are women;
Some—well, anyhow they're human.[1]
Samuel Hoffenstein

The wonderful book on globalization (he calls it cosmopolitanism) by Kwame Appiah begins with a thought that is very different from mine: if a small baby born "forty thousand years ago" were by some magical means dropped off in New York City today and raised as a typical young girl would be, she would be very much the same as another child raised by the same family and "unrecognizably different from the brothers and sisters she left behind."[2] He may be neglecting the effects of evolution upon us. I suspect this girl, unless she was quite exceptional—and there have been some wonderfully exceptional people—would not adjust well to us. She would be obviously human, but she would have been more able to adjust to her ancient tribal life than our world. I think she might well feel tragically lost.

Why? What is globalization like? Why in entering this stage of development are we going to have to change some more? And how, most likely, will we change? No doubt these are horrifying questions to the many who have not yet accommodated even to feudalism, but changes we will have.

Who are the people who will lead this change? Who seems to exemplify the changes? Let me refer to Appiah again, this time not to the book but to the photograph used to illustrate the precis of the book, which appeared as an article.[3] The photograph shows women in the market in Accra, Ghana, surrounded by boxes of tomatoes, but their focus is on a computer: someone is showing them how find information about trade so that they can get a better price or sell more. They are keenly interested. I could not help wondering when I saw the photo (it has since become one of my favorites) how long it would be before some of their crop appeared in my supermarket. Is it there right now? What kind of contacts will we develop between forward-thinking people like that in the years to come? Will their grandchildren or great-grandchildren meet and deal with mine? What will be the complexities of that kind of relationship and how rewarding will it be? Our ties are not with some forty-thousand-year-olds miraculously whisked into the present, but with the people of this and succeeding generations. Thomas Paine lays out the case here in excellent fashion: "The circumstances of the world are continually changing, and the opinions of men change also. . . . That which may be thought right and found convenient in one age, may be thought wrong and found inconvenient in another. In such cases, who is to decide, the living, or the dead?"[4]

No one, he believes, has ever had the right to decide things forever. Every generation must decide things for itself. No generation can command the following generations. We change and grow and develop. I suspect Paine would have loved the picture of the market women. They embody his spirit.

What are global people like? A lot of them, like the market women, are entrepreneurs, including many early figures, such as the people who carried on the pepper trade that I mentioned in the maritime state chapter. An American, Elihu Yale, did so well at spices in general and pepper in particular that he amassed a fortune, and founded Yale University on the proceeds.

A few of these people were also philosophers. Here are a couple of interesting examples. One can find globalists in the Stoics of Rome, of whom Marcus Aurelius is the finest example. His guiding light was reason, which he understood must rule over our lesser qualities, our passions and appetites. The person, emperor or slave, who ruled himself properly was doing what needed to be done. Other things, outside things, things beyond one's control, were not important. In general, the Stoic was to hold himself aloof from others, unless "public necessity and the general good require it."[5] These causes must be causes of great importance, of more than any local or specialized interest. Says Aurelius, "mankind are under one common law; and if so, they must be fellow-citizens, and belong to one body politic . . . the whole world is but one commonwealth."[6]

The scope of his thinking was obviously far beyond his time. Almost all others then thought in terms of their tribes; some thought in terms of their empire; and a tiny handful looked beyond that and saw their ties and their obligations to all humanity. Today the proportions are changing. If little of the modern world would make sense to Aurelius, that more people than ever before look beyond the boundaries of tribe and state would please him, in that grave and philosophical way that a Stoic has with his pleasure. And he would recognize that, in his words, "the world is all transformation."[7]

There were others as well in those ancient times who had this vision. The Greek playwright Sophocles was one. He was as much a philosopher as an author of dramas. People in his time went to see his plays, applauded them mightily, and apparently understood them not at all. He talked about reasoning one's way to understanding natural law, and although he believed these laws were unchanging (as did Aurelius), he did see this reasoning as something that anyone with sufficient wisdom could do, and therefore it was something common to all humanity, of far more significance than tribe or state. In the greatest of his plays, his heroine, Antigone, comes into conflict with King Creon and one of his laws. Creon, who had learned of the breaking of his law, the law of the state, did a good deal of complaining about the "stiff-necked anarchists"[8] who dared to oppose him; when he learns that Antigone did the job, she was brought to him. He asked why she would do such a thing. She replies, "That final justice that rules . . . makes no such laws. Your edict, King, was strong, but all your strength is weakness itself against the immortal unrecorded laws of God."[9] Note that it is the responsibility of the individual, not the state, to decide one's conduct. Antigone's and Sophocles' true home is among those who recognize her right of choice.

So the biggest change of all that we face, the newest change of great significance, is the change from having one's own city to acquiring a global perspective, to where the world and not the city is one's home. One can be at home in Tokyo or Baltimore or Liverpool, because whatever city one is in, it *is* a city, so one finds people who share the same perspectives and values. One can even physically live in the countryside, for, with modern communications, we are instantly in touch with the rest of the world. One *finds* friends and family. There are no strangers, for where one finds people with his or her values, one is also accustomed to finding the tribal people and the feudal people around. One moves from city to globe, recognizing the background of these lesser societies, yet one is not terribly interested any longer in being in those small, sheltered, and limited ponds of the tribe or nation.

Let's take a more structural view. Maryann C. Love, author of a popular text, says that globalization is "the interdependent infrastructure of global open economies, societies, and technologies."[10] The world is more and more "transsovereign," less and less managed by the old system of sovereign states.

So what good is globalization on that view? I suppose it strikes us, first and foremost, in economic matters. Recently I purchased a new computer: on the box it said that it had been assembled in this country—alas, in one instance, not well—and was composed of parts made in China, Mexico, Malaysia, and/or Thailand. When I found that I could not contact the modem, I called the company or, more exactly, I contacted a call center in India, where a guy named "Ron" worked with me on the problem, trying one thing and then another, until he finally had me take off the cover and reinstall the modem that the US workers had screwed up. It worked. Take a look at your own wardrobe: how much there is made in the USA, and how much comes from other states? On writing these words, I went to my closet; the first four shirts I checked were from China, Nicaragua, Egypt, and Korea. Would we want it otherwise? Would we be willing to pay the higher prices if the government were to enact tariffs or quotas to keep these goods out? Would we be willing to accept the drop in quality that this would entail? And yes, it would be a drop because we would be substantially lessening competition, and that makes many companies and workers sloppy.

Care for an example? Take the auto industry, here and abroad. Most of you cannot recall this, but I can: before the end of the Second World War, almost no foreign cars were allowed in the United States. This persisted for several years after the war. The result? American car companies had little incentive to compete, and there was little innovation; there was no quality control. Cars came off the assembly line as they had done since the 1920s, and however badly they were made, one had no real choice, as all cars were alike in this respect. The only real competition was at the level of the dealers—if you had a good dealer, he would have a big staff of mechanics who would try to fix your car before you bought it. Suddenly, with the lowering of tariffs, there was a rush of VWs and Fiats and Hondas; inefficient US makers like Studebaker and Kaiser went out of business and the remaining carmakers quickly found out the necessities of quality control at the factory. Local unions that fought quality control were bought off or saw their members' jobs disappear. To put things as an economist might, mostly the marketplace was allowed to do its job, and did it well. Consumers benefited. Of course if you were a worker making Henry Js or Hudsons you were out of luck, and if you had invested in stock in one of those companies, you lost your shirt. That, again, is the way the market is supposed to work. Owner, manager, or worker, you take the risk that the consumers want the products you make. You are not entitled to force them to buy your stuff.

But here is one of the conflicts between politics and economics: in a democracy, everyone has a vote, and if those who are threatened with a loss of investments or jobs vote to keep or to levy tariffs or quotas, they may win. Under the political system, one *may* be forced to buy their stuff. Is this efficient or logical? It certainly is an affirmation of the old saw that all politics is local! This is also to say that presently globalization is still in its infancy, that it occupies a middle ground between the old system of state power and the new system of combinations of people operating across state lines, and so it handles most matters with a compromise between the two. So while your standpoint might be the free market—that is, if you look at things from an economic point of view, you would be unabashedly for opening up—but if you are a nationalist, you want to close things down and go back to the old state sovereignty system. Compromise is often the answer, for a while at least.

Why do we have such trouble passing free trade agreements? Is it not because local politics intervenes? For instance, if you want free trade in the goods you buy at the supermarket, you face total opposition from farmers, especially in the wealthy nations, who can sell their crops in their own countries well above world market prices. They have a monopoly or near monopoly because of government tariffs and quotas. They have many millions of dollars, accumulated because of their protections, and they will spend it freely to defeat any attacks upon their monopoly. As one example, in July 2006, World Trade Organization talks broke down because the politicians, always careful about such things, knew that they could not, because of pressure back home, accommodate requests of the poor nations that the rich states reduce their tariffs and quotas, so that poor farmers in those countries could have a market for their goods. A newspaper article, in discussing the breakdown, pointed out that the United States was not about to lower its tariffs, even if the deal gave the US farmers better access to foreign markets.[11] Why not? Because of the fear of large farm organizations, which felt that their ability to exploit the American workers through subsidies was so important that they did not care about world markets. After all, it is easier to make money by not farming, or by farming less and having a guaranteed market through price supports and other devices, than to work. And what US politician is going to accuse farmers of being slackers? If one looks at Europe, the European Union is having a great deal of trouble dealing with its own ravenous farmers, who want ever more and don't seem to care who they starve in other places. The article I mentioned points out that European subsidies are double those of the United States. And if you go to the newly admitted EU countries like Poland, you find an even more reactionary attitude.[12] The majority there seems to be only interested in going back and turning inward. Some of this is probably the inevitable backlash that great and fundamental change brings with it, with old tribalist attitudes coming to the fore. In Poland, it is accompanied by turning to reactionary parties, like the League for Polish Families and an aggressive anti-Jewish spirit. This is very clearly local politics for us.

How can you compromise between the economic and the political aspects of situations like these? If globalization is indeed the new wave, sweeping politics and sovereignty back as it encourages people to work, think, and invest smarter, how can we handle globalization so that there will not be a serious, violent backlash that will endanger our progress?

To start from the principally noneconomic side of things, if one goes into government work, one might take on a job with an international court helping to prosecute the people so immured in the very worst aspects of tribalism that they engage in ethnic cleansing to try and purify their state. One might also work for the United Nations Secretariat, whose workers must have loyalty to the UN, not any national government, a practice that began under the old League of Nations.[13] Curiously, often people quite supportive of the organization have trouble understanding this point and think, for example, of a person from Accra or Sao Paulo going to the UN headquarters in New York as somehow having to "live in an alien environment."[14] To a committed globalist, there is nothing alien about a city. It may take a different language to work there, it may require some shift of perspective, but home is a city where one's globalist attitude is understood, respected, and commonplace, or at least commonplace compared to the way it might be regarded in the fields and forests of a rural society.

One might also take the position of bolstering organizations such as the European Union, which originated after the

Second World War, when many states in Europe came to the conclusion that killing each other was not going to work and that continental cooperation would be better. These states—at first just six—formed what became known as the Common Market, and then, with the Treaty of Maastricht in 1993, the European Union. From time to time other European nations have joined, and the early idea of an economic trading area grew and became the dream of a united Europe. They have their own currency, the euro, although not all states use it as yet, and with the addition of Eastern European states in 2004, we have the EU marking the end of Russian hegemony in Europe.

The states range from quite large, like France or Germany, to very small, like Luxembourg. What began as just an economic tool has become a useful device in many fields. From the bureaucratic organization in which there was little interest in involving the people, the European Union now has its own democratically elected parliament. While this parliament lacks some powers, it is growing in this area quite determinedly. Europe is becoming more internationalist, less tied to particular states. As a case of internationalist democracy, the EU is remarkable. If one is a citizen of one state and happens to be living in another, one may vote for the members of parliament there and even run for a parliamentary position in that other state. It has its share of problems, such as that overindulgence in agricultural subsidies and, recently, a popular reaction against the speed of integration as shown in the rejection of a proposed European constitution. Yet that union is a very important, ongoing part of European life. The EU is already, and is becoming more of a globalist influence upon Europe and the world. Some of its nations, notably conservative Poland and England, hold back, while those like Germany, Belgium, and Slovenia plunge ahead.

One can understand that the globalization of attitudes in the government field—the sloughing off of tribal and state loyalties in favor of what is a higher, more advanced plane—can only be seen as a diminution of state control in general, the control that exhibits itself in tariffs and other kinds of blockages, the barring of businesses, workers, and ideas from outside the tribal/feudal state. This economic rather than political perspective is even more a feature of the dealings of businesspeople in such international organizations as the World Bank.

The kind of organization that is not so obviously seen at first are the nonprofit associations, groups quaintly called the "nongovernmental organizations," or NGOs, most of which have little to do with international economics but everything to do with many other aspects of global ties and concerns. As one example among hundreds, take Reporters sans frontières (RSF).[15] The purpose of this group is to provide people everywhere with clear and accurate information about what is happening both in their own area and around the world. As they say, "imprisoning or killing a journalist is like eliminating a key witness and threatens everyone's right to be informed."[16] They clearly understand the interconnectedness of the world and try to provide publicity and public condemnation for acts of violence that restrict everyone's right to know. Almost all of the interferences with our right to know come from states, although nonstate terrorists can blow reporters up just as easily as they kill small children. On the day I checked the site to discover where the problems were that day, they were citing Cuba, Argentina, China, Congo, and Ukraine for various violations. It does not matter if you are a reporter born in the United States, jailed for investigating conditions in the Congo, or a reporter born in the Congo, working in Iraq, and being shot at by US troops or executed by Shia militia there. This organization is only eighteen years old, but it has more than one hundred correspondents, who cover just about every place in the world. And there are hundreds of such groups in almost every field, interest, and specialization, most of them very young.

Global organization is getting more and more important every day, and as it grows, as more people belong, they not only join RSF or Amnesty International and help them out, but work to extend the influence of these groups; and as that influence expands, the more people join, and so the expansion is quite unstoppable.

Older international groups such as the International Red Cross are more closely tied to governments and less likely to criticize states. For example, this organization will not comment on its inspections of prisons, nor on any torture activity there, no matter how horrific. In its inspection at Guantanamo in 2004, as case in point, the committee found tactics "tantamount to torture"[17] taking place, apparently quite frequently. While the Red Cross may have remonstrated with the United States government, it would not admit to it, apparently on grounds that if it revealed such conditions existed it would jeopardize future relations with the United States or other states. This report was leaked. The most the organization would say is that it "does not publicly discuss its findings . . . The Red Cross Movement supports the US government in its efforts to comply with the Geneva Conventions."[18] They meet with mixed success in their efforts, perhaps because of their deference to states. The torture at Guantanamo, according to the leaked report, was "an intentional system of cruel, unusual and degrading treatment and a form of torture,"[19] and this conduct has evidentially continued, as the United States has ignored the report. Note that in their once-secret report, the Red Cross said the torture system was "intentional,"[20] while publicly they were willing to pretend that the United States was trying to abide by the rules.

One result of that attitude has come out in the case of the suicides reported at that place. Three Guantanamo prisoners hanged themselves rather than endure the treatment any longer. The commander at the concentration camp, Admiral

Harry B. Harris Jr., offered a most amazing comment, indicative I believe of the official position: the three who committed suicide, he said, "have no regard for life, neither ours nor their own. I believe this was not an act of desperation, but an act of asymmetrical warfare waged against us."[21] One wonders how the admiral would react if he were imprisoned without end and tortured, or how pleased he would be if he attempted suicide by starvation, only to have his masters strap him to a metal chair, ram a tube down his throat, and shove food down it? The Red Cross has not commented, of course, but other groups, notably the Center for Constitutional Rights, have made known their vigorous opposition, and even before this, coalitions of government groups such as the UN panel investigating the matter and the Council of Europe have insisted that Guantanamo needs to be shut down and the United States must substantially improve its treatments of its captives, all of which the United States has refused to do.[22] But the new United States Army field manual specifically rejects torture and inhumane treatment of all detainees under their control, even if they also accept the politically imposed mandate that the words "Geneva Conventions" not be mentioned.[23] Of course this will not apply to those who are held by the CIA in foreign countries, who will continue to be tortured.

Reactionary states such as the United States, Russia, and Cuba (all of which imprison at least some of their own dissidents) may slow the growth of these international rights groups, and they may restrict their activities within their state, but there seems little doubt that over the long run, state influence or control over these groups will shrink to the vanishing point. But these states, and the people who run them and who support them, do raise a question: how far can the global people go in pursuing their goals of peace, freedom, and economic development, since so many people today are unable and unwilling to go along with them? Problems are raised, objections are flaunted, angry fists are shaken by demonstrators, and ballots against change are cast in democracies, while in Baghdad, tribes kill each other without respite. It seems like a terrible time. Are things really so bad? I think not. But before I get to that, let me answer some of the most common complaints about globalization.

The broadest and most common complaint is that globalization does a lot worse by poor people and poor countries than by rich people and rich countries.

Take as first exhibit the state of Malawi, and the complaints, taken mainly out of Love, that is, Professor Love, who I quoted earlier regarding the definition of globalization.[24] Her positions in this matter are these: a great many people in Malawi die of AIDS. Among other reasons, she complains, is that they die because US citizens are only taxed seventy-five cents a year to treat AIDS and other diseases. She further objects that the World Trade Organization rules have limited sending drugs for this condition to Malawi because Malawi cannot pay for them. Malawi is also too poor to make generic equivalents.

She also complains about a legitimacy gap,[25] that is, she questions if such international government organizations really represent the poor, or if they just represent the bankers to whom Malawi owes much money. And last she complains about the market system—this being where she claims there is an ethical gap. Drug companies care only about the money, not about human life, or more broadly, she believes that the market ignores human suffering: it may even thrive upon it.[26]

Let's look at it another way. HIV, as we know, is mostly preventable. One does not even have to abstain from sex, which is by far the quickest and simplest means of transmission. That is good, because we know that abstention is simply not going to happen, because like all other species that reproduce sexually, we have powerful sex drives. We have those drives to ensure that our species survives. People, deer, turtles, chickens, all of us reproduce up to the limit the environment will bear. Many species have predators that prey upon them, as wolves prey upon deer. As the deer increase their population, there is more food for wolves, so they increase until such point as the wolves kill so many deer that the deer population cannot support the wolves, and so a sufficient number of the wolves die off or do not reproduce, and so on. We humans also have historically reproduced in that same pattern. We had families and hoped that we produced enough offspring so that they would be able to support us in our old age. Also we find that we have regularly been subject to predators, mostly members of our own species.

But when industrialization occurs, production of goods rises, and we do not need so many children, so the sex drive is partially diverted by custom, as has become apparent in the European Union, where prosperity has brought lower rates of birth, even in the states most recently admitted to that organization.[27] Customs and methods develop to avoid or terminate pregnancy when the costs—monetary, social, and otherwise—of child rearing outweigh the benefits. Most devices to prevent pregnancy are, like condoms, simple, inexpensive, easy to use, and extremely effective. Condoms are every bit as useful in preventing AIDS as in preventing pregnancy, and they are a lot less expensive than AIDS drugs. But many churches worldwide oppose condom use, and these groups make serious efforts to ban condoms as immoral. Their stance is evil and absurd. People are not going to stop having sex; nor, in most places, especially the poorest, are people going to have only one partner their whole life, for there people die of all manner of diseases some rarely found in wealthier states. I note further that according to the CIA factbook, an excellent, if limited basic source for quick information about a country, that out of a population of almost twelve million in Malawi, almost one million have AIDS. There are few old people, less than 3 percent of the population is over sixty-five, and the average age is just over sixteen. The average life expectancy is under thirty-eight years, and the average woman gives birth to more than six children.[28]

I further note that 20 percent of the population is Catholic, an equal percent are Muslim, and almost all the remainder are Protestant, of whom almost all belong to one or another very fundamentalist church. Along comes Professor Love, a professor at the Catholic University of America, who blames drug companies for the problems of AIDS in Malawi. She needs to look closer to home. When her church and all the like-minded fundamentalists of all religions, not just Christians, are willing to accept the idea that people may legitimately defend themselves against AIDS by the use of condoms, and when they go to Malawi and hand out at least as many condoms as religious pamphlets, then they may gain some moral ground on which to criticize others.[29] Do we want to spend thousands of dollars per person to treat individual cases of AIDS, let millions of people suffer needlessly and horribly, and then die of the disease anyway? Do we want this to continue endlessly into the future? Or do we want to stop it now, or at least stop it as much as any other viral disease is ever stopped? Is it wrong to blame the American taxpayer for his or her stinginess or the World Trade Organization for bureaucratic ineptitude and wrongdoing? Granted, the latter can be blamed just fine in other areas.

But how about blaming the bankers, in particular the people at the World Bank? After all, the country owes just over three billion dollars, a staggering amount for that country. Why do they owe so much? Individual banks were eager to loan it, because the World Bank would guarantee the loan. If, as is currently being explored, the debt is forgiven, the banks that made the foolish loans will suffer no loss, since the United States will simply take three billion out of the pockets of its taxpayers and give it to these banks.

Some of you may remember back in the 1980s when the

US government paid off a lot of the debt of Mexico and simply reimbursed the banks by the same method. So I guess these loans are not foolish after all. If Mexico or Malawi pay them off, the banks get their money back plus interest; but if not, they just get it back from the American state. What have they to lose? And by the way, if you ask what the money goes for, it goes mostly for paying off politicians. Politicians get fine homes and Swiss bank accounts; but people hoping to improve the infrastructure of the country get very little.[30] Can we realistically place all the fault at the doorstep, or at the teller's window, of the banks?

One other path here needs mention, and I apologize from digressing from this talk of disease. Be patient with me, we will return to it in a bit. This other path provides a case where the World Bank can be partially faulted. Malawi is a thin, landlocked country. It snakes through the middle of Africa. It borders one major lake, Lake Nyasa. This state is about the size of Pennsylvania, although a rather long and skinny Pennsylvania. Ninety percent of the work force is in agriculture. At best, cities there are small, starved for capital because it is stolen by the political system, which otherwise hates or ignores them. What is the principal crop? Tobacco. Why? Given the declining importance of tobacco as a world crop and the damage it does to the soil (the country is too poor to fertilize properly) where will their income be in a few years? It is only six hundred dollars a year per farmer right now. Banks give loans to develop tobacco because ignorant farmers don't know how to farm anything else. It makes one rather ambivalent about banks, even if banks are not responsible for giving lessons on farming and conservation.

Finally, one more note about poor Malawi. There is most clearly one area in which their predicament is not their fault. Malawi is a former British colony, having been known as the British Protectorate of Nyasaland before 1964. They were drained by their masters for a century or more. No useful ideas or institutions were left behind when the British left. Moreover, that whole area of Africa was divided up according to the convenience of the European colonial administrations. Parts of nine major tribes were dumped into Malawi. As a result, the first leader there ruled as a dictator on his authority alone. The country did not even have a common tribal background to develop from, so the people will need a long time to progress from tribalism and resume their growth.

Let me move back to one last point about disease. As globalism spreads, diseases can be spread more quickly than ever before. Marburg, Ebola, and many others can break loose; and if the virus that causes the flu in birds can change in just one way so that it can be easily spread by one human to another, it could cause a global pandemic. If we now have a good vaccine for people for this disease,[31] it will not be so good if the virus mutates in some unexpected way, or if a new and even more dangerous threat appears.

Of course these things have happened before. It can call to mind the time of the Black Death in Europe, or the spread of disease from America to Europe, and vice versa, during the days when Europe discovered a new world. No doubt many people then were amazed that disease could spread to them so quickly.

Today the spread is even faster. Suppose, for example, the bird flu virus evolves to a form where people can readily catch it. A person picks it up, for instance, in Angola. The person gets on a plane and sneezes, spreading it to most of the other passengers and crew. The plane makes three stops, letting off passengers at each, before the passenger disembarks. Many of those whom he has infected spread it in their own cities. One of them in, say, Marseilles, sneezes while walking through the terminal, and the clerk at the coffee shop catches it and she spreads it to her family. Her child takes it to school, and spreads it there. Others do the same at their stops. The passenger ends his flight in London, and he will spread it there. A steward he has infected has a flight to New York in the morning, so he spreads it across the ocean. His son gets it, passes it to his mom. She is standing next to you at the grocery store the next day, and you pick it up from her. And you bring it to one of my classes and infect all of us!

The threat of disease is great, and it far more likely to affect more people than all the bombs of terrorists. The threat can come from anywhere. Case in point: in 2004, the College of American Pathologists, which helps labs test themselves and their abilities, requested Meridian Bioscience of Cincinnati to prepare test kits. Among the things to be tested for was a Type A influenza. "The lab selected from its stockpile the deadly 1957 H2N2 strain."[32] That's correct, "deadly." Labs throughout the United States, and to a lesser extent in other countries, possess between them all the deadly types of germs and viruses that we know. Many have been kept alive for generations. Why? The reasoning is that just in case they make a comeback, we would have them in the lab already, so we can whip up a vaccine much more quickly. Some of the most deadly are held in the facilities run by the US government. In reality these bioterror facilities exist to allow the state to use biological weapons during wartime in both offensive and defensive ways.[33]

Anyway, one of the labs being tested, the National Microbial Lab in Canada, detected the strain in March 2005, and traced it back to the kit they had gotten. They notified the World Health Organization (WHO), which notified all the labs involved in eighteen countries, and urged them to destroy the samples. Dr. Klaus Stohr, influenza chief at WHO, said, "The risk is relatively low that a lab worker will get sick, but a large number of labs got it. And if one does get infected, the risk of severe illness is high, and this virus has shown to be fully transmissible."[34] The labs destroyed their all-too-realistic test kits, and fortunately no one was infected.

So is globalization at fault? Should we close our borders? That would be to cut off most of the remedies we have to effect a cure. One article starts off by pointing out that "the growth of megacities in all countries . . . creates conditions for the spread of infectious disease. Megacities, with their extremely dense populations, facilitate epidemic outbreaks."[35] It continues by pointing to the incredibly unsanitary conditions in which many people live in slums in the large cities. The authors also point to remedies, however, such as those sponsored by the World Health Organization, as partial responses. But the authors regard these partial remedies as not nearly enough, and they conclude that globalization could be "one of the casualties" of the rapid spread of disease,[36] diseases they believe that globalization would help spread.

They are quite correct that the forces of reaction could temporarily prevail and shut borders, sending world trade smashing into a worldwide recession and starving to death hundreds of thousands of people. In some states that would work, where people are not allowed to know the benefits of their state shrinking and their society growing stronger. But more and more people are impelled to worldwide cooperation, fewer and fewer to narrowness and isolationism.

Moreover, the growth of large cities creates conditions for the containment of disease just as quickly as it leads to the spread of disease. Large nearby hospitals with specialists and good staffing cannot be maintained in rural areas without ruinously overburdening people's wallets. My wife and I lived for several years in rural Kentucky before we moved to urban New Jersey. One of the greatest differences we noticed after the move was the improvement in medical care. How much better is the care in urban New Jersey than rural Bolivia? This is just one of the many ways that it is helpful to have more people living closer together in industrial cities.

Disease starts and breeds in poor environments, and rural

environments are poorer than their urban counterparts. If one looks at the people who have so far contracted bird flu, one finds that they have gotten it mostly from chickens, and they are the people who live on chicken farms directly with the birds. To stop this disease, we need to go to the farms. If we look at mad cow disease, it starts on the farm, where vile conditions create it. To stop mad cow disease, we must teach farmers not to feed animal by-products to their cows. And if we want to take care of these things, we will have to consult medical scientists and practitioners who received their medical training in the cities.

In effect, urbanization, while it is not exactly the same as globalization, is at least a cousin. Better medicine and cleaner habits protect us against disease far better than closed borders and fearful cowering in the countryside. The city is cleaner and has far better health practices than rural areas, especially including those nastiest of rural areas, the slums that grow up around cities because their inhabitants are forbidden entrance.

Finally on disease, a brief note on polio. Polio cripples and kills; it does so because of ignorance, backed by fear and sometimes violence. If everyone gets inoculated, polio will be gone. How does ignorance work against us? Example: In 2003, several states in Nigeria stopped vaccinating children. Ignorance (a politician there) claimed that the vaccine was intended to give people AIDS or was part of a Christian plot to sterilize Muslim women, so many people stopped inoculating their children. People then developed polio and spread it within two years to people in eighteen other states in the country.[37] In Bareilly, India, when volunteers come with the vaccine, people run in fear, crying to their fellow villagers to hide their children. Again, they believe the city people are trying to poison them or sterilize them.[38] In their ignorance, many parents simply will not care enough to think to open the door. If their children are crippled or die, what concern is it of theirs? Far more important to have protected themselves against the curse of the city. It is hard to find a case where ignorance and death are so closely related.

An equally headline-grabbing issue is immigration (we have to talk about it regularly, it seems), and here, the complainers about globalization seem to reverse themselves and find that change and development hurt rich countries more than poor ones. Shut the borders through fear, and please make the whole rest of the world go away, says the citizen of the wealthier state, I don't want to look at reality any more. This problem is just as evident in the United States and other developed places as it is in Africa, where wars and famines drive people from state to state in hopes for peace and food. One may wonder if the immigrants, who seem to be called *refugees*, if they are fleeing from something like war and thereby deserving of some help, or *immigrants*, if they are moving to get a job and thereby deserving nothing, have been artificially and unrealistically classified. But is not fleeing extreme poverty in hopes of not starving a bit like fleeing war in hopes of avoiding being killed and tortured? We have set up these artificial lines or borders—sometimes originally set by colonial powers, as in Africa; sometimes set by the results of war, as between Mexico and the United States; sometimes set naturally, as at an ocean. When someone crosses a line, those on the side crossed to tend to get upset, unless they (or, more precisely, the state that they believe is looking out for their interests) set the crossing rules.

A perceptive column by John Tierney in the *New York Times* is useful here.[39] Tierney points out that in the 1950s, a flood of immigrants came into this country illegally from Mexico. Over half a million of them were caught in 1951 and the number rose later to a million. It seems astounding to think that fewer than two thousand immigration agents caught that many. One wonders how many entered and stayed successfully. But the numbers fell, and by 1955 the immigration authorities felt they had secured the border and the rate of illegal entrance continued to fall, down to only forty-five thousand in 1959. What happened? The United States enacted a "guest worker" program, and thousands of workers were allowed to enter legally. The few who continued to come in illegally were mostly criminals. What happened was that temporarily the US state recognized that people who wanted to come and work were an asset and should be welcomed. After all, the virtues we have talked about as conducive to growth and development were and are exemplified in these people. Essentially these workers were told that they had to jump some legal hurdles that the state set up, and then they could cross the border and work and benefit themselves, their families, those they were working for, the customers for their products, and their new communities. They were willing to do so, and so the crime of illegal immigration ceased for the most part in being a crime, and it ceased being a problem.

In time, the people of fear in the United States gained the upper hand again. Some of these people feared that they would lose their jobs to these workers, and rather than work harder, they wanted to outlaw competition. Others feared that these workers were being "exploited," which, to them, meant working harder than they themselves wanted to, or working for a lower wage than they would choose, and so they would prefer that these "exploited" workers be forced to go back home and starve. That way they would be out of the sight of the delicate people who fear working so much. Others feared that the workers were different. They were of a slightly different shade or color, they practiced a different religion, they spoke a different language, and so these people of fear—whose parents and grandparents who were of various shades and hues, who spoke many different languages and practiced different religions—wanted to bar them, because the country might be changed by them, just as the country had been changed by their own parents and grandparents. "Stop change after it benefits me," the fearful people cry, and they vote for equally fearful people in the legislature who pass laws making coming into the country illegal again. And yes, I guess we are back with the thought that immigrants have cooties.

Tierney offers a useful note at the end of his essay: He quotes a comment by an immigration official in 1958, who was asked in a congressional hearing how his agency could control the border if legal immigration was ended. He said that they couldn't: "We can't do the impossible, Mr. Congressman."

Actually, it is possible to stop immigration, at least for a while. One simply uses the methods used by East Germany during the Cold War to keep their citizens from out-migrating: they put up walls and used guns to kill anyone who tried to get over the walls. The United States can do the same to immigrants. This state can kill as many immigrants as try to come in, and that may stem the tide. And we can round up the ones already here. The legal groundwork for this has been established: in June 2006, a federal court judge, John Gleeson, ruled "that the government has wide latitude under immigration law to detain noncitizens on the basis of religion, race or national origin, and to hold them indefinitely without explanation."[40] Note that Judge Gleeson does not bother with the Constitution in pursuit of a way to assuage his fear; ignoring the Constitution, which in Article Five says "no person"—not, "no citizen"—can be subject to this kind of rounding up and detention, the judge sees only through the eyes of fear and prejudice. It will be left to more advanced people to overrule this decision.

These laws and rulings that deprive people of their livelihood and their freedom because of their differences are indeed evil in the precise fashion I have discussed: they are not adapted to conditions, that is, they are not adapted to reality.

They hurt rather than help; they destroy rather than develop. For the complexity of today, with all the varieties of cooperation and competition, all the shades of relationships we now see expanding before us, are changes that are literally life giving to many. Laws and rulings that crimp and handicap all of us, not just the noncitizens in our midst, laws and rulings that attempt to prevent our adjusting to our lives and our circumstances and needs, will not last. They are the bitter-end attempts by people who either cannot think of what will benefit all of us, or are not willing to think. In either case, they will disappear and be replaced by laws and rulings more in tune with people's needs, just as the guards and machine guns at the Berlin wall are gone, indeed, just as the wall itself, the East German state, and the ideology that inspired them, are all now part of history.

A final note on this matter: the description of this dispute applies to all states, and the idea of which state is rich and which is poor is relative to the point of the view of the refugee/immigrant. Take the island of Dominica; it is divided into the Dominican Republic and Haiti. The Dominican Republic is rich compared to Haiti. The island has a long history of hate and vicious tribal behavior. For example, in the 1960s, the Dominican dictator, Trujillo, had many Haitians in his country (estimates vary between 18,000 to twice that many) killed in a Caribbean version of ethnic cleansing.[41] The claim was that the Haitians were slightly darker in skin color than the Dominicans.

Anyway, lots of Haitians continue to come across the border to the Dominican Republic, because they are so poor, and would like to earn money working in the tobacco fields. The response? As the mayor of the Dominican city of Guatapanal, Jose Perez, said, "There's too many Haitians. If the government is not going to help us get rid of them, then we will do it ourselves."[42] Another Dominican, one who worked to oversee the Haitian workers, said, "They are people who do not use bathrooms. . . . They walk around drinking and making a lot of noise at night. Sometimes the men dance with each other."[43] Cooties again! Of course there is a great deal of corruption in the Dominican Republic, so the same police who track down the workers are often the same police who are bribed to let them in or let them stay. When interviewed, one of the Haitian workers said that he and his fellows had simply come to avoid starvation, and that "we do all the work, but we have no rights . . . we have to stay hidden in the shadows."[44] They stay hidden so they will not be sent back to starve to death, or be killed on the spot.

Is this repression not tribal behavior? That worker is of another tribe and therefore he has a different appearance and dirty habits. He is not really human, so we should beat him up and send him back where he belongs. Maybe we should kill him. Bash and Trash, the barbarian twins, are loose in our world, doing their best to make it unsafe for more realistic people.

While all the specific reasons that people bring up for opposing some kind of opening up of the world seem so varied that it is hard to keep track of them, they all trace back to a common feeling, the feeling of fear of change. Yet reaction has never won in the long run. The tribes or states in which it has seemed to win are no longer with us. If we do not change, our tribe dies or our state dies, and the members die with it, except the smart or lucky few (or the evolved few) who get out from under the destruction. Our real option is not whether we want to change or not—it is whether we want to live or not.

So what are some of the ways we have started to change? What more positive approach can we take to illustrate which way we are heading, and which examples can help show us how we can adapt ourselves more efficiently so that we can be part of the advance of humanity, rather than its anchor, dragging along the bottom and digging up muck as others try to sail on?

Let me first look at a few things that are relevant here, starting with the values, the virtues, the ethics of society. We have talked about these principles when we talked about the rise of cities, and how closely the bourgeois values were linked with moral and economic growth. With that development, the globalist believes that entire world is his or her city, so let me reinforce the understanding of the virtues we find. They are the usual ones such as hard and honest work, a "day's work for a day's pay" and vice versa, or fairness and respect for others. We may add an emphasis on such things as Thomas Friedman lists: the openness to "change, new technology, and equality for women."[45] Friedman also develops a point that I mentioned in talking about city people, that global people are a lot less likely to go to war because war is extremely counterproductive. War is a political activity, not an economic one, and the workers and capitalists who are "imbedded in a major global supply chain"[46] understand that they don't have time to run off and kill others if they have to keep up with producing the goods and services that others need, want, and are willing to buy or exchange for other goods.

Tied in very closely with this is the value of pluralism, which, in effect, Friedman has just been talking about in an economic context. Seen from another angle, it is what Appiah calls "fallibilism,"[47] by which he appears to mean two things. One is that people understand that they often do not understand, that they could be wrong, that more evidence could turn up to make them change their minds. The other is that we are particularly fallible as regards other people. Different people have different interests, goals, and motives. We are quite different, one from another, and therefore there is no reason to suppose that one idea or one way of life will fit everyone. Therefore, the fallibilist is willing to let others make choices in their own lives just as he or she expects others to do the same for them. Those who assume their own way is the only way will have to accept that is not so, and in any event, cannot be enforced on others.

Moreover, this assumes that the political leaders of a country where a great deal of globalist economic activity takes place (if we can go back to the economy for a moment) are amenable to the influence of these workers and capitalists—that the rulers usually operate in a more or less democratic system—and that these leaders will do what in fact will benefit their people. In effect, these politicians have become corrupted, or partially so, by the development of the economic means: killing and suppressing others has lost some of its allure. Butchering others is not a good way to get votes and therefore power from a globally inclined majority of voters. This globalist corruption of tribal virtues, this impurity, as Appiah would call it,[48] seems to be very good for us indeed! Pluralism and tolerance, vices in the tribe, are now virtues.

As I have had occasion to comment before, political leaders seldom do anything from the same motives that inspire the pure in heart; their motives infrequently stand up to careful scrutiny from a jury of moralists. But when they do the right thing from the "wrong" motives, nevertheless the right thing gets done, which is the important thing in this practical world in which we live.

This can raise some questions about the role of many states in today's world. China is an example. China is booming because of its global involvement and investments, its workers in factories who produce goods shipped overseas to almost all the countries of the world and to all the major cities. China is no democracy—so how trustworthy are the Chinese political leaders? The answer is unclear. The best we can say at present is that apparently these leaders understand that if they want to maintain their power and their wealth they must allow the globalization of their country to continue unimpeded. If they should suddenly and irrationally decide to interfere in a major way, they would be depriving millions of their people of improvement, and what is even more significant, their dream of continuing improvement. The resultant explosion would make the Tiananmen Square uprising of a few years ago seem unmentionable in the same context; they would be risking everything, and it is a reasonable guess that these rulers are fairly prudent individuals. Having few of the values and virtues of productive people, they nevertheless understand how important these beliefs are to their own benefits. They are rather in the position of those feudal lords and barons who first encouraged the new deal on their manors, and saw the wealth they reaped for themselves.

An interesting contrast was recently provided when two Chinese women looked back: Nie Yuanzi, whose poster may have ignited the "cultural revolution" in 1966, which nearly destroyed the economy and killed many hundreds of thousands, said recently, "We believed the party was great and graceful and correct, and you were obedient to the party."[49] But she went back on her loyalty, tried to quit the party, and lives now by the grace and help of friends. Now she says, "Democracy should be really promoted so that each person can express their opinions about state affairs . . . even if an opinion is not correct, it must be allowed."[50] The other woman, Wang Rongfen, saw the activities of the state in Tiananmen Square in the 1960s and wrote a letter to Chairman Mao, which got her a life sentence to prison, a sentence that was removed only after Mao died. The "cultural revolution," she had told him, "is one man with a gun manipulating the population."[51] What better way to punish truth than to imprison the one who speaks it? But realistically, how ineffective are these efforts to destroy change?

The specific lesson we can draw is that the older generation of Chinese politicians were not interested in helping their people—their rhetoric was only a covering for their reactionary hatred and their love for killing and power, their feudal/tribal ideas. The new leaders are equally uncaring, but they are capable of learning, and so far they seem to have done well. If they stumble, if they forget that they are in "power" only if they do not abuse their positions overmuch, then they will not have those positions any more, and the Chinese will move on to democracy, which will provide a better way to guard against such conduct in the future.

Another point of interest here was provided recently when Qiu Huadong talked about the "new elite" there, who work in China or without, who constitute a kind of advanced group of "international freeman"[52] who are leading the way for globalization in China. Note that Oppenheimer has foreseen the development of people like this, when he talks about the "freeman's citizenship,"[53] which he sees occurring in the future.

One might contrast this with the current Russian leadership, which apparently has not learned the lessons of their Chinese counterparts. In Russia, the start of democracy that people like Mikhail Gorbachev and Boris Yeltsin tried to bring about has been pushed under the heel of the boots of the old secret police, the KGB, and their leader, and now the leader of Russia, Vladimir Putin. The government runs rampant and corruption is the order of the day, a corruption in no way challengeable by law, because, in effect, Putin is the law. There are no basic freedoms such as speech, press, or property. As a result the country is in turmoil and the economy is in a tailspin. People starve, and the bureaucrats laugh and live it up in their villas just as in the old days of communism in the Soviet Union. A survey in 2005 by Transparency International[54] (another one of those NGOs that is keeping international pressure on states today) reveals that Russia ranks with states like Albania for corruption. One commenter says, "at least the uniformed extortionists have squeezed out the organized bandits."[55] Even that rather back-handed optimism seems wrong, because while by a wide margin people believe that the police are the most corrupt in Russia, since the police take bribes from the terrorists, Chechens and others, terrorism nonetheless continues to claim lives. An even more recent report by INDEM, the Russian think tank, said that the equivalent of $37 billion in bribes are paid every year in that state, " a sum that is roughly equivalent to the revenue portion of the 2002 federal budget and equals about 12 percent of the country's gross domestic product."[56] It is clear that Russia has few of the virtues needed for having the economies of city life, much less those needed for entering the world of globalization. One of the few bright spots there is that organizations such as INDEM still function. But if Russia is going to play catch-up with the world and join in our globalist voyage, it has a long way to go.

On the other hand, in extraordinary contrast, India is developing with a speed and in ways that almost no one could have imagined just a few years ago. The exceptionally prescient individual here is John W. Chapman, who as long ago as 1969 was writing that "the boldest of all extensions of hopes is to India."[57] He assigned increasing pluralism and voluntary association, building from and superceding the caste system, as reasons why this would happen. I suppose that most of us have had some contact with Indian growth, especially in the area of the call centers. Surprising? Not really. The call centers in India work because the people there *work*; they work quickly and efficiently, they understand their new values, and their state does not do much to prevent them. Someone may claim these workers are also paid low wages, and the proper response is, if that was the key, the call centers would be in Mongolia or Haiti or Russia. Of course, pay is a consideration, but it will not give a job to someone who will not learn how to do the work. There is still a great deal of corruption in India; there is a great deal of poverty, as well. But the people have the right virtues or they are rapidly learning them, and that is the key to success. Thus expansion continues, with the new development being that many factories are moving to India to make use of the talent. True, in some rural areas where expansion has not yet appeared, the corruption of the state apparatus is still strong; these are the areas where the Maoists still roam, as we have seen.

What we are seeing here is the increasing complexity that globalization brings us. Cities can be complicated, but the whole world is even more so. The global development is one that takes the virtues of the city and expands upon them. The globalist individual cooperates with like-minded people in many countries. It is sometimes thought that this is capitalism, and that capitalism involves some sort of dog-eat-dog competition between ruthless billionaires. The people who believe this may well be tribalists at heart, who see the very limited and basic cooperation within a tribe and do not notice that tribes compete with each other in struggles that will wind up with one tribe exterminating the other or killing off most of the members of the losing tribe and enslaving the pitiful remainder.

The bourgeois values are those that involve peaceful interaction with others for mutual benefit. People running companies do not kill their customers, they cooperate with them out of mutual interest for mutual satisfaction. Saddam Hussein, Kim Jong Il, and George W. Bush are heads or former heads of states, not entrepreneurs. Nuclear and biological weapons were ordered to be developed by states. And it was the dictator Idi Amin who ate people, not Bill Gates.[58] Such competition as exists in cities and globalist environments exists more than ever under the control of cooperative actions, agents, and groups. The increasing complexity of our world can have it no other way. Crony capitalism and special benefits are fading away.

I close my comments here with two brief notes. The first exemplifies the difference between the tribal outlook and the global outlook. When the war started in Lebanon in 2006, two letters to the editor on the topic appeared in the *New York*

Times. The first letter said that whatever Israel did was justified because it was defending its people. If "civilians will be victims," that was irrelevant, because killing innocent people is necessary for defense. The bad side was attacking civilians, therefore anything goes for the good side. The second letter pointed out that Israel was playing their enemy's game by their enemy's rules. Vengeance was not useful, and if Israel wanted to survive, it was going to have to be better than Hizbollah, not worse; they were going to have to figure out a way to bring peace.[59] What the letter writers say has a lot to tell about what kind of people they are, what they think and act like. It can also tell a great deal about the individuals who read the letters by what their reaction is.

The other note is from the wedding feature in the *Times*. A couple was pictured, one born in the far western plains of the United States, the other in Lima, Peru. They met in Peru. They are in the public health field, and have gone from place to place in this world, studying and practicing their specialties. Now they believe that they will devote their lives to "global public health,"[60] and go around the world together, supporting and helping each other and the world. I wish them success and happiness, and I hope that they are followed by many other couples and individuals in many different walks of life. Indeed, I know they will be. They are our future.

NOTES

1. From "Verses Demonstrating," in Samuel Hoffenstein, *Poems in Praise of Practically Nothing* (New York: Horace Liveright, 1928), p. 24.

2. Kwame Anthony Appiah, *Cosmopolitanism: Ethics in a World of Strangers* (New York: Norton, 2006), p. xi.

3. The photo, "High-Tech Traditional," is by Lyle Ashton Harris, *New York Times Magazine*, January 1, 2006, p. 4.

4. Thomas Paine, *The Rights of Man*, as found in *The Complete Political Works of Thomas Paine* (Chicago: Belford, Clarke, 1885), p. 24.

5. Marcus Aurelius, *Meditations*, trans. Jeremy Collier (New York: Home Book Company, n.d.), p. 58. I prefer this translation to what I regard as the stilted one by Long.

6. Ibid., p. 69.

7. Ibid., p. 68.

8. Sophocles, *Antigone*, in *The Oedipus Cycle*, trans. Dudley Fitts and Robert Fitzgerald (New York: Harcourt, Brace and Company, 1939), p. 197.

9. Ibid., p. 203.

10. Maryann Cusimano Love, ed., *Beyond Sovereignty*, 2nd ed. (Belmont, CA: Wadsworth/Thomson Learning, 2003), p. xv.

11. Tom Wright and Steven R. Weisman, "Trade Talks Fail Over an Impasse on Farm Tariffs," *New York Times*, July 25, 2006, p. A1.

12. Richard Bernstein, "After Reaching Outward, Poland Looks Back to Its Roots," *New York Times*, July 25, 2006, p. A3.

13. Lawrence Ziring, Robert Riggs, and Jack Plato, *The United Nations*, 4th ed. (Belmont, CA: Thomson Wadsworth, 2005), p. 138.

14. Ibid., p. 39.

15. http://www.rsf.org (accessed July 8, 2007).

16. Ibid.

17. "Red Cross: Guantanamo Tactics 'Tantamount to Torture,'" Reuters, November 30, 2006.

18. "Why the Red Cross Is Visiting Detainees in Guantanamo Bay," http://www.redcross.org/news/in/intlaw/guantanamo1.html (accessed June 11, 2006).

19. Reuters, "Red Cross: Guantanamo Tactics."

20. Ibid.

21. James Risen and Tim Golden, "Three Prisoners Commit Suicide at Guantanamo," *New York Times*, June 11, 2006, p. A1.

22. Ibid. They may be forced to behave better by the courts. It is also possible that political infighting may restore a balance here, the kind of maneuvering that a Madison or Adams would understand. For instance, in complaints that the executive was doing all sorts of secret

things and not admitting it, a chair of a key committee in the House of Representatives objected bitterly that he was not informed. The Constitution had been violated, he complained, but mostly he seems to have been driven to act because the secrecy was an "affront" to him and his committee. See Eric Lichtblau and Scott Shane, "Ally Told Bush Project Secrecy Might Be Illegal," *New York Times*, July 7, 2006, p. A1.

23. Julian Barnes, "Army to Use Geneva Rules for Detainees," *Los Angeles Times*, September 6, 2006, http:www.latimes.com/news/nationworld/nation/la-na-torture6sep06,0,7581942.store? (accessed July 8, 2007).

24. Love, *Beyond Sovereignty*. I make use here principally of chapter 13, "Mind the Gaps," which Professor Love authored. Note that at some points, she emphasizes the positive of globalization as well, as for example when she notes that the only constant thing about politics is that it changes, and thus she may believe that political organizations may change to handle her problems better than they do now.

25. Ibid., p. 324.

26. Ibid.

27. Elizabeth Rosenthal, "European Union's Plunging Birthrates Spread Eastward," *New York Times*, September 4, 2006, p. A3.

28. http://www.cia.gov/publications/factbook/(accessed June 2, 2006).

29. The Catholic preference for people dying of AIDS rather than using condoms continues. See Sara Kugler, "Catholics Attack NYC's Free Condoms," as reported by the Associated Press on February 15, 2007. Cardinal Egan and Bishop Dimarzio attacked the distribution of condoms, accusing the city government of New York of "blanketing our neighborhoods with condoms," which presents a rather amazing mental picture. These Catholic Church leaders want instead "what is true and what is decent" according to their own beliefs. Their own beliefs assign a much lower priority to protecting people against disease and death than a city person or a global person will accept. A number of papers carried the AP story, including the *Asbury Park Press*, which provided a cut down version on February 16, 2007, p. A14.

30. And, of course, if one remembers the arguments of Jane Jacobs in an earlier chapter, the money would not do much anyway, as this kind of development must come from people who build cities. Money from outside is generally wasted, although very small loans of one or two hundred dollars often work well.

31. Enise Grady, "Maker Calls New Bird Flu Vaccine More Effective," *New York Times*, July 27, 2006, p. A1.

32. "Deadly 1957 Strain of Flu Is Found in Lab-Test Kits," Associated Press story in *New York Times*, April 13, 2005, p. A17. Incidentally, bird flu is H5N1.

33. Ibid.

34. Ibid.

35. Dennis Pirages and Paul Runci, "Ecological Interdependence and the Spread of Infectious Disease," in Love, *Beyond Sovereignty*, p. 260.

36. Ibid., p. 263.

37. Celia W. Dugger and Donald G. McNeil Jr., "Rumor, Fear and Fatigue Hinder Final Push to End Polio," *New York Times*, March 20, 2006, p. A1.

38. Ibid.

39. John Tierney, "Securing the Border (Again)," *New York Times*, June 6, 2006, p. A21 (the op-ed) page. Succeeding quotes are from this essay.

40. Nina Bernstein, "Judge Supports Broad Powers of Detention," *New York Times*, June 15, 2006, p. B1.

41. Two places to start the search for this information are http://www.k12.ca.us/classes/social_science/latin_america/dominican_Republic.html and http://www.thediamondangle/marasco/negleg/Trujillo/html (accessed June 1, 2006).

42. Ginger Thompson, "Immigrant Laborers from Haiti Are Paid with Abuse in the Dominican Republic," *New York Times*, November 10, 2005, p. A16.

43. Ibid. Note, by the way, it is a reasonable assumption that bathrooms are pretty scarce in a Dominican tobacco field.

44. Ibid.

45. Thomas Friedman, *The World Is Flat* (New York: Farrar,

Straus and Giroux, 2005), p. 324. This book, a huge best seller, by just existing, and by selling to so many people, is an indication of how widespread the globalist population is today.

46. Ibid., p. 421.

47. Appiah, *Cosmopolitanism*, p. 144.

48. Kwame Anthony Appiah, "The Case for Contamination," *New York Times Magazine*, January 1, 2006, pp. 30ff. His article makes an even better case for impurity and mixture than does his book.

49. Howard W. French, "Hearts Still Scarred 40 Years after China's Upheaval," *New York Times*, June 10, 2006, p. A4.

50. Ibid.

51. Ibid.

52. As quoted in Howard W. French, "In China, Children of the Rich Learn Class, Minus the Struggle," *New York Times*, September 22, 2006, p. A1.

53. Franz Oppenheimer, *The State*, trans. John Gitterman, introduced by C. Hamilton, http://www.franz-oppenheimer.de (accessed June 14, 2007), p. 102, or the entire chapter 7 in his work.

54. Check out www.transparency.org. There are a number of interesting things here, such as their "Bribe Payers Index," and on the positive side, the "Integrity Awards" they give out. (Accessed October 14, 2006.)

55. "Blood Money," *Economist*, October 20, 2005, www.economist.com/cfm?story_id'5061669 (accessed October 27, 2006).

56. As reported in "Russian Corruption Study," http://www.templetonthorp.com/en/news31 (accessed October 27, 2006).

57. John W. Chapman, "Voluntary Association and the Political Theory of Pluralism," in *Voluntary Association*, Nomos XI, ed. J. Roland Pennock and John W. Chapman (New York: Atherton Press, 1969), p. 106.

58. I find it interesting to note the hypocrisy of the old state people. Professor Love attacks people who freely work together, saying "Efficiently functioning markets can still allow millions to die." Love, *Beyond Sovereignty*, p. 324. There is no free market for

death. Killing one's customers is not something that city people or global people do, for quite obvious reasons. But states can maintain the capacity to wipe all human beings off the globe, whether by spreading killer diseases or radiation, against which there is only the occasional, half-hearted protest by the conservatives and reactionaries among us.

59. "Letters," *New York Times*, July 18, 2006, p. A20.

60. "Weddings," *New York Times*, July 2, 2006, Style section, p. 7.

Part 3

THE FUTURE

CHAPTER 6

AUTONOMOUS PEOPLE

The same influences that maintain the State of War, though long since effete, will then close it, and humanity will enter a new and better period of existence, the period of Peace and Liberty.[1]
G. Molinari

Gustave de Molinari may be counted among the more hopeful of the nineteenth-century thinkers, but his is no forlorn voice in a swampy wilderness of servility and strife, either. He has had his followers and supporters who have persisted right down to the present.[2] I shall argue that they had good reason to persist, that indeed a new "period of peace and liberty" will ensue. It will be a time when the political means is eliminated and the economic means will be the norm. According to Oppenheimer, this time will be something he calls the "freeman's citizenship,"[3] a time, he feels, when exploitation will have ended and "the economic means alone will exercise sway."[4] But in suggesting that at one time people approximated this condition, he misses his own point, perhaps because he is so optimistic; in the similar way, Tom Paine suggests that much the same sort of idyllic time existed briefly at the time of the American Revolution.[5] More important however, is the point Paine adds that the more "perfect civilization is, the less occasion has it for government, because the more does it regulate its own affairs, and govern itself."[6] More and more as people change, increase the complexity of society, and develop the values we have discussed, they need less and less government. People have a "natural propensity to harmony"[7] and this produces a free society, where government no longer exists. People have always had a degree of harmony on a lesser scale. Of course, we cannot survive without some sort of cooperation with others. Man as a curiously solitary being in a constant war of all against all, in Hobbes's phrase, has never existed. Nevertheless, cooperation was restricted to members of the person's own tribe and then people were forced by governors to cooperate. It is only fairly recently that we have seen voluntary cooperation developed beyond one's tribe, and even in the present era, nations such as the former Yugoslavia were held together more by force than any sense of supratribal loyalty or even acceptance. If lonely pioneers like Paine could talk about people living together without government, it surely must have seemed visionary at the time. Today I think the idea may be broached without too many people regarding one as a mere visionary, or worst of all, a "philosopher" whose head is not quite straight compared to his or her more practical contemporaries. I think we need to start off with a few of those philosophers whose thinking really is more practical than some people might like. I'll start with Robert Wolff and finish with Herbert Spencer, with stops in between for Appiah and a few other interesting folks. But first we should probably ask what an autonomous person is, to explain the point of this chapter.

This person is someone who has gone beyond globalization, which is a condition of compromise between politics and the rest of life, a world in which politics is only weakened. During the time of globalization, the pull of the national state becomes less important than one's life in a fuller community. For the autonomous person, politics is dead, a relic of his or her past. More precisely, such a person is one who has become, to use the words of Mill, completely self-reliant and self-dependent,[8] even beyond the scope that Mill foresaw. There is no longer a doubt in the mind of such an individual that he or she is capable of handing such problems as may arise; there is no doubt that he or she, through intelligent cooperation with others, can deal with life's problems. It is the luminous lack of self-doubt that puts such persons beyond the reach of the state. It was an old saying of war resisters that "wars will cease when soldiers refuse to fight." These people will know soldiers only from old movies and antique shows on what their ancestors used to call "television." But that gets a bit ahead of the story, which here begins with Robert Paul Wolff and his interesting support for, and study of, "anarchy."[9]

Wolff begins with a description of the problem that leads to his book: he had supposed that the foremost problem of democratic political philosophy was to determine "how the moral autonomy of the individual can be made compatible with the legitimate authority of the state."[10] Basically, he admits defeat, and as a consequence, he says, he became a "philosophical anarchist."[11] I suppose we might interject here a little discussion of what anarchism is. All it really means is "without a leader." We interpret that to mean "without a government," although the two are really quite different. Some people have supposed, based on the activity of a number of individuals who have called themselves anarchists, and who have fought to equalize us by bombs rather than the state, that anarchists just want to provide complete equality by force. The "anarchists" realize that we cannot be equal under current conditions as states will enforce the existing inequality (we have called it *exploitation*, as it is done by force) for the benefit of their own class, and so anarchists bomb governments on behalf of their political theory of equality.[12] So I suppose a philosophical anarchist is a non-bomb-throwing person who has serious doubts about enforcing equality or anything else. And just to complete the picture, there are people who are strongly opposed to any overall equality, yet call for an end to the state. Some of these people style themselves "anarcho-capitalists."[13]

But to Wolff: the state to him is simply a group of people who "have and exercise supreme authority within a given territory."[14] By authority, he says he means "the right to command," and the "right" this person or persons have to be obeyed by those commanded. This is not the same thing as *force*, that is, the power of the state to put a gun to one's head and force obedience. If you have seen those signs that tell us to buckle up "because it's the law," you have seen a state attempting to use authority. We will use our seatbelt *because* someone in authority has ordered us to do so. It is not the same as when we use our seatbelt because if we do not we will be arrested and fined or as when someone without a seatbelt survives an accident and then pays the state a fine and loses his or her insurance—all, one supposes, as a way for the state to tell us we have the good luck to be living. That is force or coercion. None of this will do. When we have a duty to obey, we see that the state has mandated seat belt use and that alone is sufficient to make us buckle up. The governor or attorney general or whoever has proclaimed that henceforth all citizens shall wear seatbelts when in cars, and that is enough. We have a duty to obey the governor, whether he says to do something that is to our personal benefit, or he says to not eat sugary foods or he says to double his salary. "We hear and we obey," for we confess thee as our master.

But there is a problem with this. The problem is that when we undertake a study of moral philosophy (some regimes have banned this as harmful to our health, that is, if they catch us doing it, they will kill us), we have a grave difficulty, which is that we find that people are responsible for their ideas and their actions. People have free will.[15] And if a person has free will, she or he clearly has the responsibility of deciding if a particular action is right or not, as well as the related choice as to whether he or she should engage in it. What is "right" to do? Right by whose standards? By the standards of the thinking individual, who "insists that he alone is the judge of those constraints. He may listen to the advice of others, but he makes it his own by determining for himself whether it is good advice."[16]

The requirements for the individual are substantial; not only is a person responsible for determining what it is right for him or her to do, but he or she must determine the standards by which this action is judged. In a sense, however, the responsibility is not as great as that assumed under a state's authority, where persons who are "in authority" must try to determine what it is right for others to do. One's actual responsibility begins and ends with the self.[17] Someone might wish to try to turn this into an attack upon free and voluntary cooperation. But I am simply pointing out that any person is responsible for whom he or she cooperates with, why the cooperation is undertaken, and what the results might be for oneself and others.

The person who has accepted responsibility for himself or herself and no one else is free and autonomous. If someone pretended not to be responsible for his or her own actions, but claimed "the devil (or the president) made me do it," the person might be free, but is not autonomous because no personal responsibility is acknowledged.

One is logically driven to this point: since the essential mark of the state is authority, or the right to make rules and have them obeyed because they are the rules, and the essential mark of the person is autonomy, the right to make her or his own rules and to abide by them, one of the two principles must give way, and it ought to be the state. Each individual ought to "deny that he has a duty to obey the laws of the state simply because they are the laws."[18] The individual does not obey the laws of others—he obeys his own laws, and in many ways (but by no means all), the two coincide. The autonomous individual does not refuse to commit robbery because of the laws of New Jersey or Mauritania forbid it; the individual refuses because his or her personal moral sense forbids it. And thus there can be no such thing as a rightful command or a rightful duty to obey, which means that there can be no such thing as a rightful, legitimate, state. One might possibly think of the doom of Antigone, who denied the command of her king because his law may not overcome her own moral sense. She died, because she asserted her autonomy and was punished for it. In the future, such conflicts will not exist, and if the potential for exciting drama is thereby lessened at least in one area, the ensuing harmony and peace will make up for it.

The system Wolff proposes is *unanimous direct democracy*, "the simplest form of democratic state,"[19] a form that, however, he will finally reject as too complicated for the world to effectively use. Briefly, his idea is to have a system of voting machines tied in with television sets (today we would use the Internet), and people would vote after hearing the debates of the day on various policies. We would cut out the middleman, the politician, and take direct responsibility for voting (or not, if we chose). This is an arrangement that would perhaps please everyone except the politician, who will bitterly oppose it. I remember many years ago, the city in which my university operated was undergoing a change from one form of local government to another, and my department head was in charge of the committee to draft the new system. I recall suggesting to him that he might propose direct democracy for the city, and he was aghast at the idea: his buddies, the council members, would be out of their jobs! In short, it is dubious that this could be passed, and even if left up to the people, one would have to have a great number already committed to the idea of personal responsibility, which is not likely to happen, except in isolated cases, for some time.[20]

Moreover, while this takes care of direct democracy, it does not take care of the *unanimous* part. In order for this to happen, the individual who is supposed to obey the law would be in actual fact one of those who passed the law; says Wolff, "His obligation to submit to the laws stems . . . from the fact that he himself is the source of the laws which govern him."[21] Each individual in the society who voted, voted for the bill. Hence, if one was a member of such a society and voted for, say, a bill to prevent people from forging prescriptions of a painkilling drug, which bill became law, and one were later found to have done some forging, it would pretty well say that the individual had no wiggle room. He would be guilty without reservation or room for evasion. An alternative possibility, useful perhaps for noncriminal matters, is for people to pass unanimously some principles of compulsory arbitration under which disputes between individuals might be properly settled. Then, if one is charged with something like breaking a contract, the dispute might be settled without violating anyone's right to autonomy. The principle under which one was charged will have been settled beforehand by unanimous consent.

Now all this is very well, but it is not remotely possible to be established, especially when Wolff goes on to suppose that people ought to "set private interest aside and pursue the general good."[22] One is tempted to suggest that Wolff seems overly anxious to preserve government and not very clear about how, if ever, people are going to change to bring about his preferred utopia. Wolff is more effective on the attack, when he suggests that present arrangements are no more than "elective guardianship,"[23] and that we aren't free when representatives vote independently of our wishes, nor when laws are passed when neither the representatives nor we understand what they are doing. "Nor," he points out, "can men be called free who are subject to secret decisions, based on secret data, having unannounced consequences for their well-being and their very lives."[24] This Wolff is very much worth listening to.

If Wolff's view is clearly and closely related to politics and its functioning and theory, the outlook provided by Appiah is just as clearly cultural.

Remember back in the earliest days of 2006, when it was reported that a Danish newspaper had run cartoons that were regarded by the Muslim population in that country as insulting to their prophet? Most of the Muslims, at least all those that made the news, betrayed their tribal sentiments, one saying, for example, "We are being mentally tortured."[25] On the other hand, Soren Krarup, a retired priest and leader of the Danish anti-immigrant party, pointed out that Jesus had been satirized there as well, and that the Muslims ought to learn to take their lumps. This is a perfectly logical position. If you can have a painting of Jesus with an erection, showing Muhammad with a turban shaped like a bomb seems fair enough. Unfortunately Krarup doesn't stop there, but goes on to say that "Muslim immigration is a way for Muslims to conquer us, just as they have done for the past 1,400 years."[26]

Both sides need to give it a rest and accept the fact while in the past it was acceptable to kill everyone who disagreed with you, that is no longer so. Did anyone force Muslim readers to sit down and stare at the cartoon? Did anyone force them to move to Denmark? On the other hand, if the priest thinks that the Muslims have been trying to take over Denmark for 1,400 years, he is just as absurd as the imam.

The proper response to this kind of poisonous tribal looniness is the one that Appiah proposes: the remedy of cosmopolitanism, the end of cultural purity and exclusiveness. People everywhere are learning more and more about differences, and as they do so, they are helped in getting used to one another and to the cultural differences that we find everywhere. Appiah quotes the Roman playwright Terence, who in his "The Self-Tormentor," delivers what Appiah calls the "golden rule of cosmopolitanism . . . 'I am human; nothing human is alien to me.'"[27] This does not mean that we wind up agreeing with each other. The fear that it does is based on the old "monkey see, monkey do" rule of human conduct, which is only moderately true, even for monkeys. If people adopt something that is better, why should we object? If we do object, it is only because we are still enthralled by tribalism; the way of my grandfathers is simply better, and if you don't like it, you are not a member of my tribe—you are some kind of infidel—and I will kill you. Some people believe that kind of conduct is hard-wired into us.[28] By now, I hope I have convinced you otherwise. If people are still in a tribal state, they may well have the option of ending it; people can recognize it has long since outlived its usefulness, and if some individuals now do not have the genetic influence to do so, their descendants soon will. We need to think and act for ourselves, rather than let our earliest stage of development suffice for all time; we need to suppress our brutal rages and act in a way that meets our needs today, not the needs of our remote ancestors. We need to take and learn from many cultures, mixing and matching as our needs dictate. We need, as Appiah puts it, a great deal of "contamination" of our civilization and our ideas. As far back as the culture of the maritime state, some people, as we have seen, have recognized this.

We are evolving toward a general human culture that makes hash out of primitive, monocultural preferences. Some things are simply no longer justifiable, and there are no circumstances that can justify them. Do you still want to live in a hut, make your clothes out of animal hide, and hunt and gather like your ancestors? You may live in as much filth and squalor as you like, but you may not impose your conditions on others. Do your tribal requirements call for genital mutilation of your daughter? Tough. You will not be allowed to do so, but you may do it to yourself if you wish. "Cultural difference" is no justification for mutilating or murdering. As Appiah points out, today, "In northern Nigeria, mullahs inveigh against polio vaccination while sentencing adulteresses to death by stoning. In India, thousands of wives are burned to death each year for failing to make their dowry payments. *Vive la difference*? Please."[29]

Contrast this, if you will, to the notion of complete cultural relativism advanced by philosophers like Castoriadis, who proposed (seriously, I suppose) that "we affirm that all cultures have equal rights."[30] But only *people* have rights, and to say, for example, that a woman's right to keep her clitoris intact if she chooses, must, in certain cultures, fall before the right of the priest to carve it up, is obscene. If it was acceptable by the standards of a thousand years ago, it is obscene today. Are we therefore all the same? Does the fact that, throughout the United States and many other places, a requirement and understanding that one person must not kill or mutilate another make us all the same? Of course it does not, even though there are some reactionaries today who would like to think so, and wish to demand that everyone speak English and be Christian and (the worst horrors) not sing the national anthem in anything but English—Georgia accents allowed, at least temporarily—and above all, never burn the flag! It is symbols more than reality that matters to the followers, and perhaps even their chiefs in the tribes, and while human tribes were once connected to reality, reality has long since passed them by.

Truth is, as cosmopolitans understand, there are lots of values worth living for and living by, and one cannot follow them all. Different people will, as a matter of course, adopt differing values; do we know which is best? No. You may well know what is best for you; indeed, as an adult I expect you to either know or work hard to find out, but there is no reason to suppose I know what is best for you. Mill points out that we ought to inspire one another to do better in all the aspects of one's life. He tells us not to destroy our individuality nor stifle it, but "within the limits imposed by the rights and interests of others"[31] to develop this individuality, as it will lead to all sorts of marvelous new things that were never contemplated when one undertook the original activity. When one cultivates and develops one's life, that life can "become a noble and beautiful object of contemplation . . . [by this activity] human life also becomes rich, diversified, and animating, furnishing more abundant . . . [nourishment] to high thoughts and elevating feelings, and strengthening the tie which binds every individual to the race, by making the race infinitely worth belonging to."[32] So much for the crude and mean sort of cultural diversity that proclaims the right to mutilate women and children or kill outsiders because that is the custom of one's tribe.

Changes, contaminations, adjustments, variations, alterations, modifications—all these work against purity; they work for the individual and against the tribe. And in the coming world, they will help us survive, grow, and prosper. Tie these things together and they will, within the circle of our free will, which apparently enlarges as we develop as a species, make us more worthy candidates for continuing survival. Individuals are becoming really good (that is, humane, adaptive) and wise. Another way to say this is that we are dealing with greatly increasing complexity within ourselves and in regard to our dealings with others. When people are autonomous, balances must be made every day internally, not by some external authority. Individuals will weigh complex factors and make their own decisions, which in earlier times would have been made by governors or emperors, presidents or clan leaders.

Most of those who have advocated such change are known as *social Darwinists*. Tribal, feudal, or state publicists tend to use this term as referring to mean, evil, and nastily selfish people, those who grind the heels of their boots into the face of the starving masses. This propaganda has pretty well succeeded, as *social Darwinist* is such a low term that it is normally announced in the hot and angry fashion more appropriate for attacking a child molester or a cannibal. And indeed, after the presentation of Hannibal Lecter to us in the movies, cannibals may come off a bit better than the Darwinists. I am not really sure what a social Darwinist is, aside from the obvious point that such a person apparently believes that we are mammals. Whatever it is, it is something low and disgusting, according to a great many people. It is not surprising that those who believe in progress would like being called "futurists" or some other such label instead.

Just so that I can clear the air and report almost every possible bad thing to be said about the Darwinists and their alleged ringleader, Herbert Spencer, let me use the writings of Dante Germino, who was a well-known political thinker a few years ago. Germino believes Spencer to have been a "liberal extremist," who "vitiates" liberalism's "often generous concept of liberty" by exhibiting "a harsh and narrow rationalism and by an unthinking commitment to the productivist ethos."[33] Spencer is a rationalist—quite unthinkingly so, of course—who, since he is rational, likes it when people are productive, and productivity to Germino means "spiritual death." Germino further implies that Spencer is a some sort of protofascist and that he "bastardized" utilitarianism (that is, did not keep it pure, whole, and uncontaminated); that he fouled up again by mixing together theology and biology; that he engaged in "illicit" philosophical operations, whatever that may be, and engaged in a "bogus enterprise"; was obsessed, supported a "simplistic" creed, was foolishly consistent, in opposition to common sense, had all the "qualities of the perfect ideologue," and was ruthless. Finally, he appears to accuse Spencer of plagiarizing Darwin, despite the fact that Spencer published his *Social Statics* in 1851, while Darwin did not announce his views until 1858, and did not publish until the following year.

This has not gone away. I noted in the newspaper recently that the Catholic pope, in talking about evolution, attacked "successful adaptation" as "ultimately . . . bloodthirsty."[34] In contrast to his church's record of care and loving concern, I assume. Tribalists often make up in vitriol what they lack in judgment, and Spencer's bad rep is, mostly, a bad rap. Someone who proclaims a bold new advance must put up with that.

Spencer was not actually that exciting. The media would have a hard time if he were alive today. He did not hobnob with presidents and kings and famous actresses. He did not take part in scandals or duels or any other of the doings of society. If he were a contemporary, Paris Hilton would not know him.

Undaunted, we shall push ahead with a bit of what Spencer actually said. To begin with, Spencer suggests that people have what he calls a "moral sense," using *sense* here in rather loose fashion. That is, from the beginning we have always been social animals, and in our dealings with our fellows, we have needed a feedback mechanism (the term is not his) that informs us when we are behaving properly in order to maintain the social bond. This moral sense, he believes, helps make us work in proper combination with our fellow tribesmen, city dwellers, or others, and it gives us a sense of rightness when we have acted in accord with our requirements and our surroundings.[35] We are thus impelled to right action by a biological mechanism that leads quite directly to our surviving, and surviving well, in a particular environment.[36]

An interesting and different analysis of this mechanism is found in William Irvine's new book, *On Desire*. Irvine talks about our Biological Incentive System, or BIS, which rewards us with feelings of pleasure for acting in accord with our needs of survival. He argues, however, that the incentives and disincentives of the BIS are wired into us, and that alas, the wiring remains long after its use has become harmful or even dangerous to our survival. For example, we are wired to like sweets and to eat them, necessary in an earlier period when the danger of starvation was omnipresent. Today, many of us, those of us who have left tribal and feudal settings behind, still find we are impelled to eat as if we could starve at any moment. Thus Irvine finds our feedback system faulty and in need of repair, but no one around who can fix it. Hence he hopes to find ways to circumvent it.[37]

But whether it may be in need of repair or not, and whether one chooses to follow the analysis of Spencer or pursue the ideas of Irvine, the approach indicates that we are not blank slates upon which anything may be written, but we are innately inclined to give expression to certain types of feelings and abhor others. While different people may live in many different types of societies, from the simple to the complex, these societies stimulate certain types of conduct—cooperation, if you will—to let us get along with others and let the individual and the society survive.

And in learning how to relate our feedback mechanism to our needs, reason does have a role to play—perhaps a bit more with Spencer than Irvine, but without thinking, there is no humanity. And as our reasoning may be done well or poorly, as our reasoning varies as does our perception of reality, so we can be adapted well or badly to our condition and our needs of survival. To state things a bit differently, we may or may not be adjusted to our society, just as our society may or may not be adjusted to the needs of survival. Spencer talks a good bit about the "organic" character of society, or about a "social organism," realizing that there are just some similarities or likenesses between an organism and society. One should not take the language of "social organism" literally according to Spencer, although David Sloan Wilson, a contemporary, seems to take it that way.[38] The likenesses are there, however. A society will grow; while growing, it will become richer, more diverse, more complex; it will develop increasing interdependence among the individual members. It will become recognizably a society, a collection of individuals with much in common, rather than a chance aggregation of individuals. As such, it will combine previously uncombined individuals in different ways. Individuals will affect each other in more complex ways. For instance, if a society is globalized, it needs much greater cooperation in different ways than ever before, making use of a call center in India or an international nongovernmental organization like the Red Cross, and individuals may be challenged to perform according to the expectations of their rapidly broadening society, even if they have some characteristics of people of an earlier time. But if, for example, one has a society highly attuned to music, a tone-deaf individual will not fare well. No society can pull out of people what is not in them to begin with.

So what are we looking for in a society of autonomous people? People who are fully capable of doing for themselves, who know when and who to work with, and what the conditions need to be for effective cooperation for whatever the task at hand may be. This is needful: as Spencer remarks, "Civilization [that is, autonomous society] is the last stage"[39] that we will get to. We know now that really there is no "last stage," but only the next one. But in terms of the contrast between the barbarism of the tribal and feudal stages and the ongoing transformation we have seen in the city and global stages, it may not be such a poor way of putting it. Of course, barbarism was appropriate in its time when we absolutely needed a state, and our leaders, whether kings or presidents, served a more or less useful function. The value of that function has diminished as we have become more complex beings, as we have become more like autonomous people and less like serfs and kings. We no longer need to be forced into cooperation, but as we become autonomous, we will understand better and better the requirements of peaceful cooperation. It will be a cooperation that is more complex and more intense than ever before, far more so than city life or even global life, and it will be carried out voluntarily and peacefully. The old organizations that forced us will no longer be needed; the new society will be a society without war, where conflicts are resolved without hatred and violence. Each individual, as Spencer has it, will be able "to do the best he can by his spontaneous efforts and get success or failure according to his efficiency."[40] When we are autonomous people in an autonomous society, we will have "a condition of things in which peace and order shall be maintained without force or the fear of force."[41]

We will not need force because, as autonomous people—that is, as people who are self-reliant to a degree almost unknown today—we will not be interested in the death or destruction of any people, nor will we care to use force against others to make them think or act as we want, not as they want. We will not be the fearful people we meet on the street, who are so desirous of compelling others that they cannot abide when someone thinks about religion or politics or morals or science differently than they do. "There oughta be a law" will no longer be heard. We will have produced an environment, a society, in which human beings can assert themselves in the full and dignified stature of which they are ultimately capable. This will occasion the elimination of the principal instrument of force, the government.

There was a time, back in our early tribal days, when there

was no government, just all-powerful social bonds, and under autonomy with very different social interests and at a very different level of complexity in our cooperative relations, government will disappear again. As Spencer says, "The time was when the history of a people was but the history of its government. It is otherwise now."[42] Government used to mean everything; now, more and more, social influence has taken its place. Progress is toward less government. "Constitutional forms mean this. Political freedom means this. Democracy means this."[43] Government is a temporary thing, suitable for an underdeveloped people. The more civilization we acquire, the less the use for government, the more it gets in the way. "[A]s civilization advances, does government decay."[44] To cap our spree of quotations, government "is an institution serving a purely temporary purpose, whose power, when not stolen, is at the best borrowed."[45] And Spencer's chapter 19 is titled simply and provocatively, "The Right to Ignore the State."

By now it is surely obvious why Spencer has had very bad press. The thought that power over others may some day cease to exist is a new and frightening thing to most people even in our semicivilized world. Almost everyone believes power is necessary, and when a few gangs of ruffians call themselves terrorists and kill people, the great masses of people rise up, throw overboard all civilized restraint and call for more government and more secrecy in government, and excuse any atrocity government commits on grounds that it is done to tribal enemies. Remember in *Hyperion*, it states as principle that "in this world, a man must either be anvil or hammer."[46] Many of us still believe it—we accept the dictum of Hoffer, that there is a "full savor of power . . . from the mastery of man."[47] And whichever side of Longfellow's equation we fall upon, we regard it as true. We submit or we dominate. In the past and perhaps even today, many of us have liked doing it. But more and more people are taking responsibility for themselves, and when they are stampeded by fear into following a prior and less-viable pathway, soon they return to doing things the way that their natures now intend them to act. It may be, as Hoffer argues, that it is a "diversification and distribution of power"[48] that allows or encourages this to happen, but it does now happen, more rapidly than prior generations could have imagined. It may be a bit too soon to suppose that today it is clearly in our nature to take individual control over our own lives, but it will be no surprise or exaggeration in the future.

A final subject that needs discussion here is the question of rights. This is closely tied into the matter of "human nature." A few sentences above I used the word *nature* twice to talk about human beings. Is there such a thing as human nature? Yes, but it is not the "human nature" we commonly think of, fixed for once and all. While the tribesperson of many thousands of years ago—or the tribesperson of today—is clearly and closely related to the autonomous person, they are not the same. They have many different traits. Their genetic backgrounds are different. As a student of mine, Dara Elaine Trust, proposed, our nature is "a mutable, evolutionary aspect of our species."[49] To put it another way, if there is *a* human nature, it is an ultimate existence toward which we are developing. We have yet to realize our nature, although sometime in some unguessable future, we may do so. Let me take this up in connection with rights.

John Locke, the brilliant English philosopher and politician, can start us on our way. We have above all, he believes, a right to property, that is, to our own selves, our body, our minds, and to the things that our bodies and minds create. If one labors on a piece of land, one creates value: one mixes the soil with one's efforts, and this gives one the right of ownership. If one admits the right of self-ownership, one finds all rights based on this. Thus, for example, one may speak freely on one's own property, or in an assembly hall one has rented for the occasion; but one cannot reasonably claim the right of free speech standing on the neighbor's front porch. To do so would deny the neighbor the right to his or her self, and to the things that he or she has labored for. One may send one's words whizzing around the world electronically, but one has no right to demand that another read them.

Based on this kind of analysis, the conventional wisdom would go on to say that if one has a right to the front porch, therefore one has a right to protect that porch against interlopers, and therefore (there are many steps I omit here in the interest of brevity) one may use a state to protect the porch—in effect, to use a state to protect oneself. But the key question that the autonomous person asks is, "Why should anyone want to commit some evil action by invading someone else's property? How strange. I would not do that nor would my friends. Is it a case of mental illness? Is the poor invader in the grip of some dementia where he or she believes the porch belongs to them? Or is the person who is doing that thing some sort of throwback to an earlier time?" One sees here the idea of Spencer (and Wilson, in a different way) that evil lies in not being adapted to our needs. The point that Spencer makes, and that Tom Paine makes well before him, is properly taken. People who are autonomous are self-reliant; they cooperate well with others and enjoy that cooperation, not simply because it works to their own advantage, but because it works to everyone's advantage. We live in our own sphere and do not intrude upon others because we recognize their persons and their rights. Moreover, when Spencer is talking about this condition of humankind, he says, in part:

> To the primary requisite that each shall be able to get complete happiness without diminishing the happiness of the rest, we must now add the secondary one that each shall be capable of receiving happiness from the happiness of the rest. Compliance with this requisite implies *positive beneficence*.[50]

So how about rights? Rights, I suspect, are inherent in all of us. Our circumstances and our development call them out and make them manifest in the world. As we develop some scraps of self-reliance, perhaps even the desires to be self-reliant, we find that rights make sense, and we want them. Even as in the days of feudalism, when rights developed crudely and out of the interests of the ruling class, people used those rights and expanded on them. When cities began, the people there found rights to be eminently practical and effective, so they worked for more. As more and more autonomous people come along, rights will expand greatly. Autonomous people need that expansion now, and even the rest of us will benefit from the fuller realization of rights.

John Locke probably said it best in the related case of children. First he points out that "paternal power" should really be understood as "parental power," for father and mother share it jointly.[51] Next, one's rights are to do as one wishes, without interfering with the rights of others, and "without being subjected to the Will or Authority of any other Man."[52] The child is not immediately in this condition of full possession of rights, as he or she is too young to cope with the world. We do not expect a six-year-old, for example, to provide his or her own food and shelter. But the child is, says Locke, "born to" this condition.[53] Parents have an obligation to the children they bring into the world, to see to it that they are preserved, nourished, and educated, so that by the time they are older, they can take on and use effectively their full rights. They will know what they need to do and can do, and they will understand and accept the rules of the society in which they live, or at least understand they can go and live in another society if that is better for them, in their own mature judgment.[54] All of us have rights, but these rights develop in practical ways over long periods of time and through trial and error, through our evolutionary changes. Even the tribesman is born to these rights, even though he may not even be aware of them or reject them. He is still human, and his descendants will ultimately understand and use the rights that he, quite unknowingly, through the long and convoluted pathways our species has undertaken, has bequeathed to them.

Before I wrap this up, I want to briefly lay out one more interesting approach to our future society. This is an approach suggested to me recently from two sources: one, the Irvine book that I have already alluded to, although the possibility there is vaguely indicated, and the other, more directly expressed by a student who wrote a paper urging this approach. This approach is religious in nature and makes use of the thought of the "enlightened one," Gautama Buddha. While this may not be a direct concern of autonomy, nonetheless it is true that a practitioner of Buddhism may find that desires for the goods or persons of others diminishes with the meditative practices of the interesting combination of religion and practical psychology[55] that makes up that viewpoint. Aaron Werschulz has argued that "political autonomy is, essentially, Buddhist principles practiced in the global arena."[56] His argument moves along the lines that if we would recognize all people as not being strangers but friends and fellows, there would be no wars, and cooperation would be the norm. There are no doubt other religions that have a good deal to contribute here, insofar as they have overcome or are in the process of overcoming their fundamentalist fanatics.

It is hard to conclude this matter, since it is so speculative to begin with. It is, in its own way, the story of the growth of human freedom and responsibility, the tale or our continuing maturation as a species. It is the story of our species, which begins with people living in tribes, then growing into the complexities of the feudal or the maritime state. From there, people can become city folk, then global, always expanding their horizons. With autonomous society, the emphasis retains its globalist character, but the people overcome the last remnants of local interest and focus at last fully on autonomy. Politics is ended, and people begin to live in the flowering of what we might think of as the first great human civilization, where one is no longer a hammer or an anvil. It is an age of extreme individualism and unprecedented cooperation, an age where there is no fear of others, but one becomes an enormously sophisticated part of a society that yet recognizes the values of each individual participant. And in the midst of that triumph, someone will rise up and cry, "The best is yet to come! Let's see what's just down the road!"

NOTES

1. G. Molinari, *The Society of To-Morrow*, trans. P. H. Lee Warner, introduced by Hodgson Pratt, appendix by Edward Atkinson (New York: G. P. Putnam's Sons, 1904), p. 171.

2. See the incredible amount of literature at http://www.praxeology.net/anacres/htm#heritage, the Web site for the Molinari institute (accessed January 12, 2007).

3. Franz Oppenheimer, *The State*, trans. John Gitterman, introduced by C. Hamilton, http://www.franz-oppenheimer.de (accessed June 14, 2007), p. 106.

4. Ibid.

5. Thomas Paine, *The Rights of Man*, as found in *The Complete Political Works of Thomas Paine* (Chicago: Belford, Clarke, 1885), p. 349.

6. Ibid., p. 350.

7. Ibid., p. 392.

8. The details of what is involved here have already been discussed in chapter 4.

9. Robert Paul Wolff, *In Defense of Anarchism* (New York: Harper & Row, 1970).

10. Ibid., p. vii.

11. Ibid., p. viii.

12. For example, see Robert LeFevre, *This Bread Is Mine* (Milwaukee, WI: American Liberty Press, 1960), p. 75, for one who upholds this view. The anarchists who agree are so numerous that one cannot name them all, but any work by Kropotkin or Bakunin will do. A few anarchists, Dorothy Day of the Catholic Worker movement as great example, are pacifists and do not bomb, but try to persuade the state to dismantle itself.

13. The Web is full of people who are, more or less, anarcho-capitalists. See, for example, http://anarcap.unanimocracy.com/, or http://jim.com.anarcho-.htm. You may believe them to be a grave danger to the republic, or harmless. Some are interesting. (Accessed July 17, 2006.)

14. Wolff, *In Defense of Anarchism*, p. 3.

15. Without getting into the enormous swamp of the free will vs. determinism debate here, let us just say that a person, as a biological unit, is physically responsible for what that biological unit does. For those who must squirm uncontrollably in their chairs at this assertion, I recommend ten laps around the track and then a hot bath. This is not the place to settle that old chestnut.

16. Wolff, *In Defense of Anarchism*, p. 13.

17. The objector might say here that this does not cover the case of the small child. If one has a four-year-old, for example, does not the parent have some responsibility for deciding on his or her behalf? The answer of course is "yes," and I suspect they will follow the Lockean advice as advanced in chapter 6 in his *Second Treatise of Government*, introduced by Peter Laslett (London: Mentor, 1965). But more on this in a bit.

18. Wolff, *In Defense of Anarchism*, p. 18.

19. Ibid., p. 22.

20. Says Mill, in respect to representative government, what perhaps would apply with even more force to direct governing, that while a people might at present be unfit for it, "it may be for their good to have had it even for a short time (although) they are unlikely long to enjoy it." John Stuart Mill, *Considerations on Representative Government*, ed. Currin V. Shields (New York: Library of Liberal Arts, 1958), p. 7.

21. Wolff, *In Defense of Anarchism*, pp. 21–22.

22. Ibid., p. 78.

23. Ibid., p. 31.

24. Ibid. See as well, "Government Asserts It Is above the Law in AT&T Case." Late last Friday night, the government filed its reply brief, providing a last round of written briefing in advance of this week's hearing in our case against AT&T for collaborating with the government's surveillance program. Finally the administration has come out and flatly said what it has hinted at throughout its arguments: that the program is above the law. The government wrote that "the court—even if it were to find unlawfulness upon in camera, ex parte review—could not then proceed to adjudicate the very question of awarding damages because to do so would confirm Plaintiffs' allegations." Essentially the government is saying that, even if the judiciary found the wholesale surveillance program was illegal after reviewing secret evidence in chambers, the Court nevertheless would be powerless to proceed. The executive has asserted that the program, which has been widely reported in every major news outlet, is still such a secret that the judiciary (a coequal branch under the Constitution) cannot acknowledge its existence by ruling against it. In short, the government asserts that AT&T and the executive can break the laws crafted by Congress, and there is nothing the judiciary can do about it. Learn more about the case and read case documents: http://www.eff.org/legal/cases/att/. The above is copied from the *Bulletin of the Electronic Frontier Foundation* 19, no. 23 (June 20, 2006). Do Wolff's words prove prophetic?

25. Dan Bilefsky, "Denmark Is Unlikely Front in Islam-West Culture War," *New York Times*, January 8, 2006, p. A3.

26. Ibid.

27. Kwame Anthony Appiah, "The Case for Contamination," *New York Times Magazine*, January 1, 2006, pp. 30ff. The quote is from page 52.

28. David Berreby, *Us and Them: Understanding Your Tribal Mind* (New York: Little, Brown and Company, 2005). See especially p. 4, "all of us are capable of *both* tribal good and tribal evil."

29. Appiah, "The Case for Contamination," p. 37.

30. See Cornelius Castoriadis, *Philosophy, Politics, Autonomy*, ed. David Ames Curtis (New York: Oxford University Press, 1991), p. 141.

31. John Stuart Mill, *On Liberty*, ed. Currin V. Shields (New York: Liberal Arts Press, 1956), p. 76.

32. Ibid.

33. Dante Germino, *Modern Western Political Thought* (Chicago: Rand McNally, 1972), p. 220. The various allegations appear between pages 256 and 260. This much venom is only equaled in public display when someone milks a cobra.

34. Ian Fisher, "Professor-Turned-Pope Leads a Seminar on Evolution," *New York Times*, September 2, 2006, p. A3. This is the leader of the church that was responsible for the Crusades, the Spanish Inquisition, and currently in the United States is trying hard to cover over the child abuse committed by its priests.

35. Herbert Spencer, *Social Statics* (London: John Chapman, 1851). All references will be made by chapter and paragraph. This work should not be confused with a later edition, which has a great deal of the most original analysis removed. It has also been published by the Robert Schalkenbach Foundation (New York, 1970). This edition has a brief preface by Francis Neilson. The quote is from the introduction: "The Doctrine of the Moral Sense." One might wish as well to note his Lemma I, in which he develops the idea that human nature is not always the same. His examples are way out of date, most disproven by a century-and-a-half of scientific advance. The mechanism by which we change, however, is mostly valid. A bit of the same approach appears to be taken by Marc D. Hauser in his recently published *Moral Minds: How Nature Designed Our Universal Sense of Right and Wrong* (New York: HarperCollins, 2006). There are problems in thinking of nature as a designer and presuming that this design is universal. I will stick with Spencer as modified.

36. For a modern take on this, in softer words than Spencer's, see Daniel Goleman, *Social Intelligence* (New York: Bantam Books, 2006).

37. William B. Irvine, *On Desire* (New York: Oxford University Press, 2006). See chapter 7 on the BIS.

38. David Sloan Wilson, *Evolution for Everyone* (New York: Delacorte Press, 2007), chap. 21, "The Egalitarian Ape," deals with this, and seems to find the differences between an organ and an organism lie on a continuum, with few clear demarcations between them.

39. Spencer, *Social Statics*, chap. 15–para. 6.

40. Ibid., 16–9.

41. Ibid., 16–8.

42. Ibid., introduction–4.

43. Ibid.

44. Ibid.

45. Ibid., 19–3.

46. Henry Wadsworth Longfellow, *Hyperion: A Romance* (Boston: Houghton, Mifflin and Company, 1893), p. 371.

47. Eric Hoffer, *The Ordeal of Change* (New York: Harper & Row, 1963), p. 123.

48. Ibid., p. 97.

49. Dara Elaine Trust, "The Nature of Human Nature," unpublished paper, April 2006, p. 2.

50. Spencer, *Social Statics*, III–2. When he talks about the cruelty of savage peoples and their "treachery and vindictiveness," XXX–2, he also talks about the graces and goodness that are to come. He goes on to insist that our evolution is goal-directed, which we now know to be untrue.

51. Locke, *Second Treatise*, VI–52 and 53.

52. Ibid., VI–54

53. Ibid., VI–55.

54. Ibid., VI–56 to 58. Needless to say, this does not mean that I find all of Locke's reasoning to be equally valid. It is not necessary to do so, and, mostly, I don't.

55. Irvine, *On Desire*, pp. 184–89.

56. Aaron Werschulz, "The Dharma of Anarchy," unpublished paper, May 2006, p. 1.

CHAPTER 7

PEOPLE AFTER AUTONOMY

"The Universe!" the Aaron cried ecstatically. "My people, *henceforth the universe is ours!*"[1]
William Tenn

The time after we have achieved autonomy will be so distant, perhaps not in time but in perceptions, that it is very difficult to think about the subject. What we can say about ourselves in that future time, when people will be so unlike us? At the risk of totally embarrassing myself in the next few years as scientific progress hurries along, let me offer a few possibilities.

An important change that will affect all the others is that we will be increasingly in control of our genetic change.[2] Remember, here I speak in the wake of autonomy, and so I speak of us as individuals, not us as a mass driven by leaders, intent on their own gratification. Just a few years ago this seemed quite "science-fictiony," just as a few years before that genetic control was associated with the centers of power in fascism and racism. It has now become mostly dependent on individual choice,[3] so ethnic cleansing of any sort is clearly more in our past than our future. A poem I ran across speaks to this: it speaks of the end to the fear of change, to the future when people making their own choices will look back on people now as we think of "Neanderthals." This is the way our species will develop: "We can manage it for ourselves, thanks. . . ."[4]

Another important change will probably be something even more extraordinary. It may well be that economics, or the problems of production and distribution of goods and services, will move outside the social realm into the personal, and economics will no longer be a crucial matter for society—perhaps in the way that people like Jim Mason or Richard Barbrook talk about the "gift economy."[5] If things are nearly costless to produce, then most items will be given away. If death can be put off for centuries and taxes will disappear along with the state in an autonomous world, it may not be much of a stretch to suppose that economic pressures will diminish to the point of nonexistence.

The most important change of all is in the way we will regard ourselves and our former selves. Will we still think of ourselves as humans or as cyborgs, that is, a mixture of human and machine?[6] Will we be so changed, with so many genetic changes and so many computer chips in us, so many artificial limbs and organs, that we will be too different to call human anymore? Will those in the future be so very different that they will look on anyone who is unchanged as today we look upon other family members of the great apes?

We need to understand that it has been centuries since all of us have been totally, naturally, human. Remember back when you read *Treasure Island,* and picture the pirate with the peg leg. He was not a fully natural, organic human. Part of one of his limbs was wooden. Think of today. When someone dies, the funeral homes have to remove bridges and implants such as pacemakers before that person is cremated. We are adding on to ourselves. In a few years, the doctor may say to you, "That liver of yours is starting to fail. Better get an artificial one put in. It will last longer and do a better job for you." "Nip and tuck" operations now will become "remodel, replace, and improve." Insert a computer chip in our brains and our thoughts will be deeper and go faster than even the best computer today, with a capacity for independent thought that will make an Einstein look backward.[7] Our species is passing away, some believe, and Garreau offers one scientist's view that the passing of "people like us" will be a wonderful thing.[8] On that basis, I suppose the old joke will be reworked so that one cyborg will ask another in derisive fashion, "Are you descended from humans on your father's side or your mother's?" But I believe the humane emotions and aspirations will remain. We have been improving ourselves in one way or another for centuries. Humanity will remain. With other recent advances in medicine, we may be able to regrow naturally any organ or part we need, so this may obviate any concern about mechanically constructed people.

In one small way I have a regret (beside the obvious one of having the maddening curtain of the future clouding my vision), and that is that people of that distant future might think poorly of us. We do that of our forbears. I know—many of my students have said of a Jefferson or a Calhoun, "Well, they were slaveholders," and then dismiss their thought and their persons as unworthy of any further consideration. I hope that my descendants do not look back upon my generation and think with a shudder, "How awful! We are related in the terrible past to such a barbarian!" Indeed, when I ran across a book in an antique store, I thought, "How I wish I could use that title. It is so apt in such a poignant way." But the title was already taken: *Recollections of a Savage*.[9]

For all the terrible things done in this time I would apologize (no, I'm not responsible, but it has been my generation), yet I know that this is only the prelude to great things. I cannot imagine what the world, or more likely the universe will be for us in millennia to come, but there will be something exciting happening. Always our species will change. It could well be that the physical changes will be so great that people such as the members of the World Transhumanist Association[10] will turn out to be mostly correct, and we will be so different that it will be hard to recognize us. But in a key respect they will be wrong, I believe: we will still be fully human, however far we travel.

Our eyes are on the stars, while our feet will rest uncomfortably in the manure of our ancestral murders, torturing, and hatred, our own past. No matter. All will be well. Humanity will see to it.

NOTES

1. William Tenn, *Of Men and Monsters* (New York: Mentor Books, 1968), pp. 250–51. As the first quote from him came at the beginning of his story and fits with the beginning of mine, this quote came from almost the end of his story and fits almost the end of mine. There is, of course, no real start or end of the story of human beings, just the beginning and the ending of what one knows or what one can imagine.

2. For many examples of just how far we have come and what is almost within our reach, see Joel Garreau, *Radical Evolution* (New York: Broadway Books, 2005), with particular attention to his three scenarios: heaven, hell, and the in-between, "prevail." The scenario I have outlined is a mixture of the two better-for-us ones, based, I hope, on reasonable expectations of what is happening to our evolutionary growth, and what we can add to it. And it includes an understanding, which I did not find in his work, that evolutionary

choice is definitely an individual matter, not something for the Department of Defense to decide.

3. For just one example among the many happening all the time, see Amy Harmon, "Couples Cull Embryos to Halt Heritage of Cancer," *New York Times*, September 3, 2006, p. A1.

4. Brian W. Aldiss, "Progression of the Species," in *Holding Your Eight Hands*, ed. Edward Lucie-Smith (Garden City, NY: Doubleday, 1969), pp. 1–2.

5. As they are quoted in Garreau, *Radical Evolution*, pp. 254–55.

6. Garreau, *Radical Evolution*, pp. 61–65. See also in http://www.KurzweilA1.net the article by Max More, "Embrace, Don't Relinquish the Future" (accessed February 3, 2007).

7. For a witty look at this view, see L. Neil Smith's *The Probability Broach* (New York: Ballantine Books, 1979).

8. Garreau, *Radical Evolution*, p. 115.

9. Edwin A. Ward, *Recollections of a Savage* (New York: Frederick Stokes, n.d.).

10. http://www.transhumanist.org (accessed June 5, 2007).

AFTERWORD

We have met the enemy and he is us.[1]
Pogo

So what have I been trying to demonstrate here? That people have demonstrably changed their behaviors from tribal times down to the present, that they will continue to exhibit altered behaviors into the future, and that these changes have been in response to the needs and requirements of the people involved. I think many of you will find that I have amassed sufficient evidence to have made my point, although more evidence would be nice. You may wish to go back to one or another of the principal sources as Oppenheimer and Jacobs, or delve more deeply into the scientific side, to view more of the evidence in particular cases, with special attention to such ideas and references as Joel Garreau provides in his *Radical Evolution*.[2]

Further, I have made the assumption that this clear change of behaviors is accompanied by an interior change, a change in the way, for example, that a global person would view, think about, and consider participating in something like a beheading, compared to the way a tribal person would handle such things. A tribal person would delight in it, if it was an "enemy" beheaded, but would vow revenge if it was a member of his or her tribe. As Chris Hedges put it, during the war in Bosnia, all sides "ignored the excesses of their own and highlighted the excesses of the other in gross distortions that fueled the war."[3] The globalist view would be a clear recognition that giving pain to others would be a wrongful thing,[4] and even the increasingly rare and narrowly focused self-defense, when it gives pain to wrongdoers, is not a good thing in itself, just not as bad as letting them succeed. Pain is not wrong to a tribesperson. It is an ever-present event. Even in the household, the husband may beat his wife, his wife may beat the children, and the children may beat the dog. And when one loses one's tribe and is yet only adapted to that life, one turns one's life over to some charismatic leader who promises pain. As I was deciding how to present this—it was painful to me, but I did not do it because of that—I saw a picture in the newspaper of a row of Iranian women, tightly wrapped so only their eyes and hands were visible, dressed as suicide bombers, offering up the fascist salute and themselves and their children (it does not matter if one's children die, after all, any more than if they themselves die), all to the cause of their sect and their leader.[5] City or global people value themselves and children, whether their own or others, much too strongly for that. Such people have self-esteem, rather than love of tribe, or love for a tribe that has been displaced. There is none of the hatred for all who are not of the tribe or who are not approved by one's leader.

A further assumption is that a major part of this change,

both in conduct and consideration, is substantially related—causally related—to genetic change. This assumption I leave to scientists to prove or disprove, but I think I have made a reasonable case that it ought to be examined; indeed, it is in fact being examined.[6] The assumption is appropriate if for no other reason than that it is the only explanation I can find for the changes overtaking us, that make global people or autonomous people so very different from tribal people, if equally human. Is it simpler to assume that each person has such capacity for personal alteration, that each person can move from one stage to another simply by an act of will, that someone who acts and thinks globally can suddenly, with only modest motivation, turn into a pack animal or a mindless member of an angry mob, or to assume that our choices as a species are changing over time because people are changing? I think the real difference here is that someone who believes in the unlimited range of behaviors open to everyone probably must also believe in some kind of special creation of humans, that we are unrelated to all the other species on Earth, while someone who sees our choices as limited and changing respects our natural origins.

So what happens to the idea that we have free will? It requires us to say simply that while we have free will, it is not an uncaused and irresponsible will detached from all other parts of us. I have sufficient free will to imagine myself becoming a seagull and flying through air, or figure out an ingenious system of reproduction among aliens that requires three genders, but I cannot by any exercise of will become a seagull or an alien. My choices in actions are limited to what I, as a person, can physically and mentally do within my own genetic parameters. John Stuart Mill, in talking about government, puts it very nicely: Within limits, within the condition, for example, that people are willing to do what needs to be done to keep a particular form of state going, people may create and maintain that form. But if people are willing only to obey their tribal leader, they are not ready to maintain a feudal system; or if people are not ready to take the time to learn about politics and cast an informed vote, they are not ready for democracy.[7]

Free will, however, is not the subject of the book. Others have discussed it at much greater length, so I need not provide here more than this notice to indicate the bare bones of my position, just as I have indicated the bare bones of my position on scientific matters, and my invitation and encouragement to geneticists and other scientists to explore the subject further.

Remember that we are not talking about some kind of plan that evolution has for us. Evolution does not choose, *people* choose. And a process does not have a plan. The process works by random selection, changes continually occurring quite randomly in the genetic makeup of people and chimps, watermelons and dandelions, amoebas and viruses. And if we talk about such things as the Biological Incentive System with Irvine, we miss such points as Kurzweil makes, that we are not only changing, but the rate of change is accelerating.[8]

If there is or was ever such a thing as BIS, it has evidently changed. What worked for tribal peoples—not, note, *ever* worked for the whole population of the earth—it does not and cannot work in anything like the same way for advanced peoples. Scientists frequently talk about changes as what they call "heritable mutations."[9] These mutations can be caused by all sorts of things such as infections, living on top of depleted uranium shell casings, and so on. On the whole, these changes are damaging—they interfere with our abilities to deal with our environment. But sometimes, very rarely, these changes work out. They produce a change that makes it easier for us to deal with our world, to adapt better. And in these cases, that change, if it is a genuinely useful mutational change, will carry over from generation to generation. The individual who has had this mutation thrust upon him or her will pass it on, and may even have more children who are themselves a bit better adapted, or able to adapt, to the environment.

If the suggestions I have made at the start of this discussion are correct, then we are being shaped by evolution even now, and the evolutionary pattern that develops *is* indeed working on behalf of the individual and the species, though not intentionally so. There is no need to try to thwart evolution, but instead to work carefully on our own evolution, to direct our activities and our thoughts to our own improvement and to allow others to do so as they choose. We will then be able to cast aside the saying at the top of this chapter and instead say, "I have met my friends. I have recognized myself in each of them." We have a long way to go, and now that we have achieved even small abilities to help along our evolutionary process, we would be wise to use them.

NOTES

1. Walt Kelly's most famous saying, first done for an Earth Day poster in 1971. See http://www.igopogo.com/we_have_met.htm (accessed July 2, 2006).

2. Joel Garreau, *Radical Evolution* (New York: Broadway Books, 2005).

3. Chris Hedges, *War Is a Force That Gives Us Meaning* (New York: Anchor Books, 2003), p. 64. See also page 76.

4. Even then, a few globalists and autonomists are coming close to the rejection of violence of any sort, even if they may be wrongly affected by it. Their intolerance of pain may be so strong that they find it hard, if not impossible, to countenance their infliction of it, even in cases where they might otherwise have to suffer it themselves.

5. Michael Slackman, "Lebanon Conflict a Setback for Iran Reform Movement," *New York Times*, August 1, 2006, p. A9. One may suppose those involved were trying to be like good tribal men.

See also Hedges, *War Is a Force*, pp. 68–69, where he meets a man who prays that all his own sons will be chosen by God to die as martyrs for Palestine.

6. A regular review of the scientific literature is enlightening. There are good engines online (a couple of which I have already referenced) that can hook one into many of these new ideas.

7. John Stuart Mill, *Considerations on Representative Government*, ed. Currin V. Shields (New York: Library of Liberal Arts, 1958), see generally chapter 1. Note as well on page 10: "There are abundant instances in which a whole people have been eager for untried things. The amount of capacity which a people possess for doing new things and adapting themselves to new circumstances is itself one of the elements for our consideration." Mill wrote most of this in 1860. I suppose he would be pleased to find himself vindicated and more people more ready than ever to do new things a century and a half later.

8. Ray Kurzweil, "Does Evolution Select for Faster Evolvers?" January 31, 2007, http://www.kurzweilai.net/news/newsprintable.html?id'6355 (accessed February 1, 2007).

9. For a very elementary view, see "How Evolution Occurs," http://www.evolutionhappens.net (accessed July 2, 2006). It will lead you to more sophisticated places.

INDEX